ISBN 978-1-330-53192-1
PIBN 10074752

1 MONTH OF
FREE
READING

at
www.ForgottenBooks.com

By purchasing this book you are eligible for one month membership to ForgottenBooks.com, giving you unlimited access to our entire collection of over 1,000,000 titles via our web site and mobile apps.

To claim your free month visit:

www.forgottenbooks.com/free74752

English
Français
Deutsche
Italiano
Español
Português

www.forgottenbooks.com

Mythology Photography **Fiction**
Fishing Christianity **Art** Cooking
Essays Buddhism Freemasonry
Medicine **Biology** Music **Ancient**
Egypt Evolution Carpentry Physics
Dance Geology **Mathematics** Fitness
Shakespeare **Folklore** Yoga Marketing
Confidence Immortality Biographies
Poetry **Psychology** Witchcraft
Electronics Chemistry History **Law**
Accounting **Philosophy** Anthropology
Alchemy Drama Quantum Mechanics
Atheism Sexual Health **Ancient History**
Entrepreneurship Languages Sport
Paleontology Needlework Islam
Metaphysics Investment Archaeology
Parenting Statistics Criminology
Motivational

PLATE I.

BARN SWALLOW.

CLIFF SWALLOW.   TREE SWALLOW.

BANK SWALLOW.

BIRD-LIFE

A GUIDE TO THE STUDY OF OUR COMMON BIRDS

BY

FRANK M. CHAPMAN

ASSISTANT CURATOR OF THE DEPARTMENT OF MAMMALOGY AND
ORNITHOLOGY IN THE AMERICAN MUSEUM OF NATURAL
HISTORY ; MEMBER OF THE AMERICAN
ORNITHOLOGISTS' UNION ;
AUTHOR OF HANDBOOK OF BIRDS OF EASTERN NORTH AMERICA, ETC.

*WITH SEVENTY-FIVE FULL-PAGE PLATES AND
NUMEROUS TEXT DRAWINGS*

By ERNEST SETON THOMPSON
AUTHOR OF ART ANATOMY OF ANIMALS, THE BIRDS OF MANITOBA, ETC.

TEACHERS' EDITION

NEW YORK
D. APPLETON AND COMPANY
1899

TO

Dr. J. A. ALLEN

THIS BOOK IS DEDICATED

AS A TOKEN OF RESPECT AND AFFECTION

FROM ONE WHO

FOR NINE YEARS HAS WORKED AT HIS SIDE.

YƧOTƧIH JAЯUTAИ

PREFACE.

How unusual it is to meet any one who can correctly name a dozen of our birds! One may live in the country and still know only two or three of the one hundred and fifty or more kinds of birds that may be found during the year. Nevertheless, these gay, restless creatures, both by voice and action, constantly invite our attention, and they are far too interesting and beautiful to be ignored. No one to whom Nature appeals should be without some knowledge of these, the most attractive of her animate forms.

The scientific results to be derived from the study of birds are fully realized by the naturalist. But there are other results equally important. I would have every one know of them: results that add to our pleasure in field and wood, and give fresh interest to walks that before were eventless; that quicken both ear and eye, making us hear and see where before we were deaf and blind. Then, to our surprise, we shall discover that the forests and pastures we have known all our lives are tenanted by countless feathered inhabitants whose companionship will prove a source of endless enjoyment.

I would enter a special plea for the study of birds in the schools; for the more general introduction of ornithology in natural-history courses. Frogs and cray-fish serve an excellent purpose, but we may not encounter either of them after leaving the laboratory; whereas birds not only offer excellent opportunities for

study, but are always about us, and even a slight familiarity with them will be of value long after school days
are over.

Popular interest must precede the desire for purely
technical knowledge. The following pages are not addressed to past masters in ornithology, but to those who
desire a general knowledge of bird-life and some acquaintance with our commoner birds. The opening
chapters of this book briefly define the bird, its place in
Nature and its relation to man, and outline the leading
facts in its life-history. The concluding chapters present the portraits, names, and addresses of upward of one
hundred familiar birds of eastern North America, with
such information concerning their comings and goings
as will lead, I trust, to their being found at home.

After this introduction the student may be left on
the threshold, with the assurance that his entrance to the
innermost circles of bird-life depends entirely on his own
patience and enthusiasm.

FRANK M. CHAPMAN.

AMERICAN MUSEUM OF NATURAL HISTORY,
    NEW YORK CITY, *January, 1897.*

CONTENTS.

LIST OF ILLUSTRATIONS.

Full-page Plates.

ix

LIST OF ILLUSTRATIONS.

FIGURES IN THE TEXT.

TEACHERS' PORTFOLIOS OF PLATES.

THE seventy-five plates contained in Bird-Life have been colored, and, with twenty-five new plates, are now issued unbound, in three portfolios. Each plate can therefore be used separately, thus becoming the best substitute for a specimen of the bird it represents.

The additional plates are:

The one hundred plates are placed in three portfolios, as follows:

Portfolio No. 1; thirty-two plates.

Permanent Residents and Winter Visitants. (See Appendix, pages 6, 7.)

Portfolio No. 2; thirty-four plates.

March and April migrants. (See Appendix, pages 16, 17.)

Portfolio No. 3; thirty-four plates.

May migrants, types of birds' eggs, and nests of birds photographed in nature. (See Appendix, pages 20, 21.)

BIRD-LIFE.

CHAPTER I.

THE BIRD: ITS PLACE IN NATURE AND RELATION TO MAN.

The Bird's Place in Nature.*—About thirteen thousand species of birds are known to science. The structure of many of these has been carefully studied, and all have been classified, at least provisionally. Taken as a whole, the class Aves, in which all birds are placed, is more clearly defined than any other group of the higher animals. That is, the most unlike birds are more closely allied than are the extremes among mammals, fishes, or reptiles, and all living birds possess the distinctive characters of their class.

When compared with other animals, birds are found to occupy second place in the scale of life. They stand between mammals and reptiles, and are more closely related to the latter than to the former. In fact, certain extinct birds so clearly connect living birds with reptiles, that these two classes are sometimes placed in one group—the Sauropsida.

---

* On the structure of birds read Coues's Key to North American Birds, Part II (Estes & Lauriat): Headley, The Structure and Life of Birds: Newton's Dictionary of Birds—articles, Anatomy of Birds and Fossil Birds; Martin and Moale's Handbook of Vertebrate Dissection, Part II, How to Dissect a Bird; Shufeldt's Myology of the Raven (Macmillan Co.).

2

The characters that distinguish birds from mammals on the one hand, and from reptiles on the other, are more apparent than real. Thus flight, the most striking of a bird's gifts, is shared by bats among mammals. Egg-laying is the habit of most reptiles and of three mammals (the Australian duckbill and the echidnas). But incubation by one or both of the parents is peculiar to birds, though the python is said to coil on its eggs.

Birds breathe more rapidly than either mammals or reptiles, and their pneumaticity, or power of inflating numerous air-sacs and even certain bones, is unique.

The temperature of birds ranges from 100° to 112°, while in mammals it reaches 98° to 100°, and in the comparatively cold-blooded reptiles it averages only 40°.

The skull in mammals articulates with the last vertebra (atlas) by two condyles or balls; in birds and reptiles by only one. In mammals and birds the heart has four chambers; in reptiles it has but three.

Mammals and reptiles both have teeth, a character possessed by no existing bird; but fossil birds apparently prove that early in the development of the class all birds had teeth.

Thus we might continue the comparison, finding that birds have no universal peculiarities of structure which are not present in some degree in either mammals or reptiles, until we come to their external covering. The reptile is scaled, and so is the fish; the mammal is haired, and so are some insects; but birds alone possess feathers. They are worn by every bird—a fit clothing for a body which is a marvelous combination of beauty, lightness, and strength.

There is good evidence for the belief that birds have descended from reptilian ancestors. This evidence consists of the remains of fossil birds, some of which show marked reptilian characters and, as just said, are toothed.

It is unnecessary to discuss here the relationships of the birdlike reptiles, but, as the most convincing argument in support of the theory of the reptilian descent of birds, I present a restoration of the Archæopteryx, the earliest known progenitor of the class Aves. This restoration is

Fig. 1.—Restoration of the Archæopteryx, a toothed, reptilelike bird of the Jurassic period. , (About 1/5 natural size.)

based on an examination of previous restorations in connection with a study of the excellent plates which have been published of the fossils themselves.* Two specimens have been discovered; one being now in the British Museum, the other in the Berlin Museum. They were both found in the lithographic slates of Solenhofen, in Bavaria, a formation of the Jurassic period, and, together, furnish the more important details of the structure of this reptilelike bird.

This restoration, therefore, while doubtless inaccurate

---

* For recent papers on the Archæopteryx see Natural Science (Macmillan Co.), vols. v–viii.

in minor points, is still near enough to the truth to give a correct idea of this extraordinary bird's appearance.

The Archæopteryx was about the size of a Crow. Its long, feathered tail is supposed to have acted as an aëroplane, assisting in the support of the bird while it was in the air, but its power of flight was doubtless limited. It was arboreal and probably never descended to the earth, but climbed about the branches of trees, using its large, hooked fingers in passing from limb to limb.

The wanderings of this almost quadrupedal creature must necessarily have been limited, but its winged descendants of to-day are more generally distributed than are any other animals.* They roam the earth from pole to pole; they are equally at home on a wave-washed coral reef or in an arid desert, amid arctic snows or in the shades of a tropical forest. This is due not alone to their powers of flight but to their adaptability to varying conditions of life. Although, as I have said, birds are more closely related among themselves than are the members of either of the other higher groups of animals, and all birds agree in possessing the more important distinguishing characters of their class, yet they show a wide range of variation in structure.

This, in most instances, is closely related to habits,

---

* On the distribution of animals read Allen, The Geographical Distribution of North American Mammals, Bulletin of the American Museum of Natural History, New York city, iv, 1892, pp. 199–244; four maps. Allen, The Geographical Origin and Distribution of North American Birds considered in Relation to Faunal Areas of North America, The Auk (New York city), x, 1893, pp. 97–150; two maps. Merriam, The Geographic Distribution of Life in North America, with Special Reference to Mammalia, Proceedings of the Biological Society of Washington, vii, 1892, pp. 1–64; one map. Merriam, Laws of Temperature Control of the Geographic Distribution of Terrestrial Animals and Plants, National Geographic Magazine (Washington), vi, 1894, pp. 229–238; three maps.

which in birds are doubtless more varied than in any of the other higher animals. Some birds, like Penguins, are so aquatic that they are practically helpless on land. Their wings are too small to support them in the air, but they fly under water with great rapidity, and might be termed feathered porpoises. Others, like the Ostrich, are terrestrial, and can neither fly nor swim. Others still, like the Frigate Birds, are aërial. Their small feet are of use only in perching, and their home is in the air.

If now we should compare specimens of Penguins, Ostriches, and Frigate-birds with each other, and with such widely different forms as Hummingbirds, Woodpeckers, Parrots, and others, we would realize still more clearly the remarkable amount of variation shown by birds. This great difference in form is accompanied by a corresponding variation in habit, making possible, as before remarked, the wide distribution of birds, which, together with their size and abundance, renders them of incalculable importance to man. Their economic value, however, may be more properly spoken of under

The Relation of Birds to Man.—The relation of birds to man is threefold—the scientific, the economic, and the æsthetic. No animals form more profitable subjects for the scientist than birds. The embryologist, the morphologist, and the systematist, the philosophic naturalist and the psychologist, all may find in them exhaustless material for study. It is not my purpose, however, to speak here of the science of ornithology. Let us learn something of the bird in its haunts before taking it to the laboratory. The living bird can not fail to attract us; the dead bird—voiceless, motionless—we will leave for future dissection.

The economic value of birds to man lies in the service they render in preventing the undue increase of insects, in devouring small rodents, in destroying the seeds of harmful plants, and in acting as scavengers.

Leading entomologists estimate that insects cause an annual loss of at least two hundred million dollars to the agricultural interests of the United States. The statement seems incredible, but is based upon reliable statistics. This, of course, does not include the damage done to ornamental shrubbery, shade and forest trees. But if insects are the natural enemies of vegetation, birds are the natural enemies of insects. Consider for a moment what the birds are doing for us any summer day, when insects are so abundant that the hum of their united voices becomes an almost inherent part of the atmosphere.

In the air Swallows and Swifts are coursing rapidly to and fro, ever in pursuit of the insects which constitute their sole food. When they retire, the Nighthawks and Whip-poor-wills will take up the chase, catching moths and other nocturnal insects which would escape day-flying birds. The Flycatchers lie in wait, darting from ambush at passing prey, and with a suggestive click of the bill returning to their post. The Warblers, light, active creatures, flutter about the terminal foliage, and with almost the skill of a Hummingbird pick insects from leaf or blossom. The Vireos patiently explore the under sides of leaves and odd nooks and corners to see that no skulker escapes. The Woodpeckers, Nuthatches, and Creepers attend to the tree trunks and limbs, examining carefully each inch of bark for insects' eggs and larvæ, or excavating for the ants and borers they hear at work within. On the ground the hunt is continued by the Thrushes, Sparrows, and other birds, who feed upon the innumerable forms of terrestrial insects. Few places in which insects exist are neglected; even some species which pass their earlier stages or entire lives in the water are preyed upon by aquatic birds.

Birds digest their food so rapidly, that it is difficult to estimate from the contents of a bird's stomach at a given time how much it eats during the day. The stomach of a Yellow-billed Cuckoo, shot at six o'clock in the morning, contained the partially digested remains of forty-three tent caterpillars, but how many it would have eaten before night no one can say.

Mr. E. H. Forbush, Ornithologist of the Board of Agriculture of Massachusetts, states that the stomachs of four Chickadees contained one thousand and twenty-eight eggs of the cankerworm. The stomachs of four other birds of the same species contained about six hundred eggs and one hundred and five female moths of the cankerworm. The average number of eggs found in twenty of these moths was one hundred and eighty-five; and as it is estimated that a Chickadee may eat thirty female cankerworm moths per day during the twenty-five days which these moths crawl up trees, it follows that in this period each Chickadee would destroy one hundred and thirty-eight thousand seven hundred and fifty eggs of this noxious insect.

Professor Forbes, Director of the Illinois State Laboratory of Natural History, found one hundred and seventy-five larvæ of *Bibio*—a fly which in the larval stage feeds on the roots of grass—in the stomach of a single Robin, and the intestine contained probably as many more.

Many additional cases could be cited, showing the intimate relation of birds to insect-life, and emphasizing the necessity of protecting and encouraging these little-appreciated allies of the agriculturist.

The service rendered man by birds in killing the small rodents so destructive to crops is performed by Hawks and Owls—birds the uninformed farmer considers his enemies. The truth is that, with two exceptions, the Sharp-shinned and Cooper's Hawk, all our commoner Hawks and Owls are beneficial. In his exhaustive study of the foods of these birds Dr. A. K. Fisher, Assistant Ornithologist of the United States Department of Agriculture, has found that ninety per cent of the food of the Red-shouldered Hawk, commonly called "Chicken Hawk" or "Hen Hawk," consists of injurious mammals and insects, while two hundred castings of the Barn Owl contained the skulls of four hundred and fifty-four small mammals, no less than two hundred and twenty-five of these being skulls of the destructive field or meadow mouse.

Still, these birds are not only not protected, but in some States a price is actually set upon their heads! Dr. C. Hart Merriam, Ornithologist and Mammalogist of the United States Department of Agriculture, has estimated that in offering a bounty on Hawks and Owls, which resulted in the killing of over one hundred thousand of these birds, the State of Pennsylvania sustained a loss of nearly four million dollars in one year and a half!

As destroyers of the seeds of harmful plants, the good done by birds can not be overestimated. From late fall to early spring, seeds form the only food of many birds, and every keeper of cage-birds can realize how many a bird may eat in a day. Thus, while the Chickadees, Nuthatches, Woodpeckers, and some other winter birds are ridding the trees of myriads of insects' eggs and larvæ, the granivorous birds are reaping a crop of seeds which, if left to germinate, would cause a heavy loss to our agricultural interests.

As scavengers we understand that certain birds are of value to us, and therefore we protect them. Thus the Vultures or Buzzards of the South are protected both by law and public sentiment, and as a result they are not only exceedingly abundant, but remarkably tame. But we do not realize that Gulls and some other water birds are also beneficial as scavengers in eating refuse which, if left floating on the water, would often be cast ashore to decay. Dr. George F. Gaumer, of Yucatan, tells me that the killing of immense numbers of Herons and other littoral birds in Yucatan has been followed by an increase in human mortality among the inhabitants of the coast, which he is assured is a direct result of the destruction of birds that formerly assisted in keeping the beaches and bayous free from decaying animal matter.

Lack of space forbids an adequate treatment of this subject, but reference to the works and papers mentioned below * will support the statement that, if we were deprived of the services of birds, the earth would soon become uninhabitable.

Nevertheless, the feathered protectors of our farms and gardens, plains and forests, require so little encouragement from us—indeed, ask only tolerance—that we accept their services much as we do the air we breathe. We may be in debt to them past reckoning, and still be unaware of their existence.

But to appreciate the beauty of form and plumage of

---

* Notes on the Nature of the Food of the Birds of Nebraska, by S. Aughey; First Annual Report of the United States Entomological Commission for the Year 1877, Appendix ii, pp. 13–62. The Food of Birds, by S. A. Forbes; Bulletin No. 3, Illinois State Laboratory of Natural History, 1880, pp. 80–148. The Regulative Action of Birds upon Insect Oscillations, by S. A. Forbes, ibid., Bulletin No. 6, 1883, pp. 3–32. Economic Relations of Wisconsin Birds, by F. H. King; Wisconsin Geological Survey, vol. i, 1882, pp. 441–610. Report on the Birds of Pennsylvania, with Special Reference to the Food Habits, based on over Four Thousand Stomach Examinations, by B. H. Warren; Harrisburg, E. K. Meyers, State Printer, large 8vo, pp. 434, plates 100. The English Sparrow in North America, especially in its Relation to Agriculture, prepared under the Direction of C. Hart Merriam, by Walter B. Barrows; Bulletin No. 1, Division of Economic Ornithology and Mammalogy of the United States Department of Agricul-

birds, their grace of motion and musical powers, we must know them. Then, too, we will be attracted by their high mental development, or what I have elsewhere spoken of as "their human attributes. Man exhibits hardly a trait which he will not find reflected in the life of a bird. Love, hate; courage, fear; anger, pleasure; vanity, modesty; virtue, vice; constancy, fickleness; generosity, selfishness; wit, curiosity, memory, reason—we may find them all exhibited in the lives of birds. Birds have thus become symbolic of certain human characteristics, and the more common species are so interwoven in our art and literature that by name at least they are known to all of us."

The sight of a bird or the sound of its voice is at all times an event of such significance to me, a source of such unfailing pleasure, that when I go afield with those to whom birds are strangers, I am deeply impressed by the comparative barrenness of their world, for they live in ignorance of the great store of enjoyment which might be theirs for the asking.

I count each day memorable that brought me a new friend among the birds. It was an event to be recorded in detail. A creature which, up to that moment, existed

---

ture, 1889. The Hawks and Owls of the United States in their Relation to Agriculture, prepared under the Direction of C. Hart Merriam, by A. K. Fisher; Bulletin No. 3, ibid., 1893. The Common Crow of the United States, by Walter B. Barrows and E. A. Schwarz; Bulletin No. 6, ibid., 1895. Preliminary Report on the Food of Woodpeckers, by F. E. L. Beal; Bulletin No. 7, ibid., 1895. (See also other papers on the food of birds in the Annual Report and Yearbook of the United States Department of Agriculture.) Birds as Protectors of Orchards, by E. H. Forbush; Bulletin No. 3, Massachusetts State Board of Agriculture, 1895, pp. 20–32. The Crow in Massachusetts, by E. H. Forbush; Bulletin No. 4, ibid., 1896. How Birds affect the Farm and Garden, by Florence A. Merriam; reprinted from "Forest and Stream," 1896, 16mo, pp. 31. Price, 5 cents.

for me only as a name, now became an inhabitant of my
woods, a part of my life. With what a new interest I
got down my books again, eagerly reading every item
concerning this new friend; its travels, habits, and notes;
comparing the observations of others with what were
now my own!

The study of birds is not restricted to any special season. Some species are always with us. Long after the
leaves have fallen and the fields are bare and brown,
when insect voices are hushed, and even some mammals
are sleeping their winter sleep, the cheery Juncos flit
about our doorstep, the White-throats twitter cozily from
the evergreens, Tree Sparrows chatter gayly over their
breakfast of seeds, and Crows are calling from the woods.
Birds are the only living creatures to be seen; what a
sense of companionship their presence gives; how deso-
late the earth would seem without them!

The ease with which we may become familiar with
these feathered neighbors of ours robs ignorance of all
excuses. Once aware of their existence, and we shall see
a bird in every bush and find the heavens their pathway.
One moment we may admire their beauty of plumage,
the next marvel at the ease and grace with which they
dash by us or circle high overhead.

But birds will appeal to us most strongly through
their songs. When your ears are attuned to the music
of birds, your world will be transformed. Birds' songs
are the most eloquent of Nature's voices: the gay carol of
the Grosbeak in the morning, the dreamy, midday call
of the Pewee, the vesper hymn of the Thrush, the clang-
ing of Geese in the springtime, the farewell of the Bluebird in the fall—how clearly each one expresses the sentiment of the hour or season!

Having learned a bird's language, you experience an
increased feeling of comradeship with it. You may even

share its emotions as you learn the significance of its
notes. No one can listen to the song of the Mockingbird
without being in some way affected; but in how many
hearts does the *tink* of the night-flying Bobolink find a
response? I never hear it without wishing the brave
little traveler Godspeed on his long journey.

As time passes you will find that the songs of birds
bring a constantly increasing pleasure. This is the result
of association. The places and people that make our
world are ever changing; the present slips from us with
growing rapidity, but the birds are ever with us.

The Robin singing so cheerily outside my window
sings not for himself alone, but for hundreds of Robins I
have known at other times and places. His song recalls
a March evening, warm with the promise of spring; May
mornings, when all the world seemed to ring with the
voices of birds; June days, when cherries were ripening;
the winter sunlit forests of Florida, and even the snow-
capped summit of glorious Popocatepetl. And so it is
with other birds. We may, it is true, have known them
for years, but they have not changed, and their familiar
notes and appearance encourage the pleasant self-delusion
that we too are the same.

The slender saplings of earlier years now give wide-
spreading shade, the scrubby pasture lot has become a
dense woodland. Boyhood's friends are boys no longer,
and, worst of all, there has appeared another generation
of boys whose presence is discouraging proof that for us
youth has past. Then some May morning we hear the
Wood Thrush sing. Has he, too, changed? Not one
note, and as his silvery voice rings through the woods
we are young again. No fountain of youth could be
more potent. A hundred incidents of the long ago become as real as those of yesterday. And here we have
the secret of youth in age which every venerable natural-

ist I have ever met has convincingly illustrated. I could name nearly a dozen, living and dead, whom it has been my valued privilege to know. All had passed the allotted threescore and ten, and some were over fourscore. The friends and associates of their earlier days had passed away, and one might imagine that they had no interest in life and were simply waiting for the end.

But these veterans were old in years only. Their hearts were young. The earth was fair; plants still bloomed, and birds sang for them. There was no idle waiting here; the days were all too short. With what boyish ardor they told of some recent discovery; what inspiration there was in their enthusiasm!

So I say to you, if you would reap the purest pleasures of youth, manhood, and old age, go to the birds and through them be brought within the ennobling influences of Nature.

CHAPTER II.

THE LIVING BIRD.

Factors of Evolution.—If while in the fields we observe birds with an appreciative eye, we shall soon be impressed with the great diversity shown in their structure and habits. The Fish Hawk plunges from the air into the water and grasps its prey with merciless talons. The Hummingbird daintily probes a flower. The Woodpecker climbs an upright trunk, props itself with its stiff, pointed tail-feathers, while with its chisel-shaped bill it excavates a grub and then impales it with its spearlike tongue. These birds tell us a wonderful story

Fig. 2.—End of spearlike tongue of Pileated Woodpecker. (Much enlarged.)

of adaptation to the conditions of life, and, knowing that they have descended from a common ancestor, we ask, " Why do they now differ so widely from one another ? " Biologists the world over are trying to satisfactorily answer this question, and it is impossible for me to even mention here all the theories which they have advanced. However, some knowledge of the most important ones is essential if you would study the relation between the bird and its haunts and habits. The Darwin-Wallace theory of Natural Selection, in more or less modified forms, is accepted by most naturalists. As originally presented, it assumed that the continued existence of any animal depended upon its adaptation to its manner of life. Among a large number of individuals there is much variation in size, form, and color. Some of these variations might prove favorable, others unfavorable. Those which were favorable would give to the individual possessing them an advantage over its fellows, and, by what is termed *Natural Selection*, it would be preserved and its favorable characters transmitted to its descendants. But the less fortunate individuals, which lacked the favorable variation, would be handicapped in the race for life and be less likely to survive.

Without necessarily opposing this theory, the followers of Darwin's predecessor, Lamarck, attach more importance to the direct action of environment on the animal—that is, the influence of climate, food, and habit. The effect of the first two I will speak of in treating of color; the last we may use to illustrate the difference in these two theories by asking the question, "Is habit due to structure, or is structure the result of habit?" Has Nature, acting through natural selection, preserved those variations which would best fit a bird to occupy its place in the world, and are its habits the outcome of the characters thus acquired, or have the changes which during the ages have occurred in a bird's home, forcing it to alter its habits, been followed by some consequent change in structure, the result of use or of disuse? For my part, I answer "Yes" to both questions, and turn to our stiff-tailed, spear-tongued Woodpecker to explain my reply. I can readily understand how the shape of these tail-feathers is the result of habit, for the same or similar structure exists among many birds having no close relationship to one another, but all of which agree in their peculiar use of the tail as a prop; the Creep-

ers, Woodhewers, and Swifts, even some Finches and
the Bobolink, that use their tail to support them when
perched on swaying reeds, have the feathers more or
less pointed and stiffened. Furthermore, this is just the
result we should expect from a habit of this kind. But

FIG. 3.—Tip of tail of (*a*) Downy Woodpecker and of (*b*) Brown Creeper, to
show the pointed shape in tails of creeping birds of different families.
(Natural size.)

I do not understand how the Woodpecker's spear-tipped
tongue could have resulted from the habit of impaling
grubs, and in this case I should be inclined to regard
structure as due to a natural selection which has preserved favorable variations in the form of this organ.

I have not space to discuss this subject more fully,
but trust that enough has been said to so convince you
of the significance of habit, that when you see a bird in
the bush it will not seem a mere automaton, but in each
movement will give you evidence of a nice adjustment
to its surroundings. Remember, too, that evolution is a
thing of the present as well as of the past. We may not
be able to read the earlier pages in the history of a species,
but the record of to-day is open to us if we can learn to
interpret it.

This may be made clearer, and the importance of a
study of habit be emphasized, if I briefly outline the relation between the wings, tail, feet, and bill of birds and
the manner in which they are used. We are in the field,
not in the dissecting room; our instrument is a field glass,
not a scalpel, and in learning the functions of these four

organs we shall direct our attention to their external form rather than their internal structure.

The Wing.—Birds' wings are primarily organs of locomotion, but they are also used as weapons, as musical instruments, in expressing emotion, and they are some-

Fig. 4.—Young Hoatzin, showing use of hooked fingers in climbing. (After Lucas.)

times the seat of sexual adornment. As an organ of locomotion the wing's most primitve use is doubtless for climbing. Gallinules, for instance, have a small spur on the wrist or "bend of the wing," and the young birds use it to assist their progress among the reeds. A more striking instance of this nature is shown by that singular South American bird, the Hoatzin (*Opisthocomus cris-*

3

tatus). The *young* of this bird have well-developed claws on the thumb and first finger, and long before they can fly they use them as aids in clambering about the bushes, very much as we may imagine the Archæopteryx did. In the *adult* these claws are wanting.

Some eminently aquatic birds, as Grebes and Penguins, when on land, may use their wings as fore legs in scrambling awkwardly along; while some flightless birds, for example, the Ostrich, spread their wings when running.

But let us consider the wing in its true office, that of an organ of flight, showing its range of variation, and finally its degradation into a flightless organ. Among flying birds the spread wings measure in extent from about three inches in the smallest Hummingbird to twelve or fourteen feet

Fig. 5.—Short, rounded wing and large foot of Little Black Rail, a terrestrial bird. (³/₅ natural size.)

in the Wandering Albatross. The relation between shape of wing and style of flight is so close that if you show an ornithologist a bird's wing he can generally tell you the character of its owner's flight. The extremes are shown by the short-winged ground birds,

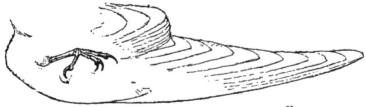

Fig. 6.—Long, pointed wing and small foot of Tree Swallow, an aërial bird. (³/₅ natural size.)

such as Rail, Quail, Grouse, certain Sparrows, etc., and long-winged birds, like the Swallows and Albatrosses. There is here a close and, for the ground-inhabiting birds, important relation between form and habit. Many terrestrial species rely on their dull, protective covering to escape observation, taking wing only when danger is so near that it is necessary for them to get under way at once. Consequently, Quail, Partridges, and Grouse, much to the amateur sportsman's discomfiture, spring from the ground as though thrown from a catapult, and reach their highest speed within a few yards of the starting point, while the Albatross is obliged to face the wind and run some distance over the ground or water before slowly lifting itself into the air. There, however, it can remain for hours or even days without once alighting.

The Frigate Bird, or Man-o'-War Bird, has a body scarcely larger than that of a chicken, but its tail is one foot and a half in length, and its wings measure seven to

Fig. 7.—Frigate Bird. (Expanse of wings, 7 to 8 feet.)

eight feet in extent. Having this enormous spread of sail, its flight is more easy and graceful than that of any living bird. I have seen hundreds of these birds floating in the air, facing the wind, without apparent change of position or the movement of a pinion, for long intervals of time.

From this extreme development of the wing as a flight-organ, let us turn to those birds who have not the power of flight. The Ostrich, Rhea, Emu, and Cassowary are familar representatives of this group. It is generally believed that these birds have lost the power of flight, and that as their wings, through disuse, became functionless, their running powers correspondingly increased. This, however, is theory, but there are birds which have become flightless through some apparently known cause. They may be found among such widely separated families as Grebes, Auks, Ducks, Rails, Gallinules, Pigeons, and Parrots.

One of the characteristic water birds of our North Atlantic coasts is the Razor-billed Auk. It is a strictly aquatic species, nearly helpless on land, which, as a rule, it visits only when nesting. Its egg is laid in the crevice of a rocky cliff, frequently at some height from the sea. During the winter it migrates southward as far as Long Island. Flight is therefore a necessary faculty, and we find the bird with well-developed wings, which it uses effectively. We can, however, imagine conditions under which it would not be necessary for the Razor-bill to fly. It might become a permanent resident of isolated islands, laying its egg on accessible beaches. Already an expert diver, obtaining its food in the water, it would not be obliged to rise into the air, and, as a result of disuse, the wings would finally become too small to support it in aërial flight, though fully answering the purpose of oars.

Apparently this is what has happened in the case of the Razor-billed Auk's relative, the flightless, extinct Great Auk. The Razor-bill is sixteen inches long and its wing measures eight inches, while the Great Auk, with a length of thirty inches, has a wing only five and three fourths inches in length. Aside from this difference in measurements these birds closely resemble each other. So far as we are familiar with the Great Auk's habits, they agreed with those of the hypothetical case I have just mentioned, and we are warranted, I think, in assuming that the bird lost the power of flight through disuse of its wings.

In antarctic seas we find the arctic Auks replaced by the Penguins, a group in which all the members are flightless. They are possessed of remarkable aquatic

FIG. 8.—Great Auk, showing relatively small wing. (Length of bird, 30 inches; of wing, 5·75 inches.)

powers, and can, it is said, outswim even fish. They nest only on isolated islands, where they are not exposed to the attack of predaceous mammals.

Among Grebes and Ducks we have illustrations of the way in which swimming birds may become temporarily flightless. With most land-inhabiting birds flight is so important a faculty that any injury to the wings is apt to result fatally. It is necessary, therefore, that the power of flight shall not be impaired. Consequently, when molting, the wing-feathers are shed slowly and symmetrically, from the middle of the wing both inwardly and outwardly; the new feathers appear so quickly that at no time are there more than two or three quills missing from either wing. But the

aquatic Grebes and Ducks, protected by the nature of
their haunts and habits, lose all their wing-feathers at
once, and are flightless until their new plumage has
grown.

It might then be supposed that permanently flightless
forms would be found among the Grebes and Ducks.
But these birds are generally migratory, or, if resident,
they usually inhabit bodies of fresh water where local
conditions or droughts may so affect the food supply that
change of residence would become necessary. However,
on Lake Titicaca, Peru, there actually is a Grebe which
has lived there long enough to have lost the use of its
wings as flight-organs.

Rails are such ground-lovers, and fly so little, that we
should expect to find flightless forms among them when
the surroundings were favorable for their development.
In New Zealand, that island of so many flightless birds,
the requirements are evidently fulfilled, and we have the
flightless Wood Hens. Here, too, lives the flightless
Gallinule, *Notornis*, and in this family of Gallinules,
birds not unlike Coots, there are at least four flightless
species inhabiting islands—one in the Moluccas, one in
Samoa, one on Tristan d'Acunha, and one on Gough
Island. The last two islands are about fifteen hundred
miles from Cape Good Hope, and have evidently never
been connected with a continent. There seems little
reason to doubt, therefore, that the ancestors of the
Gallinules now inhabiting these islands reached them
by the use of their wings, and that these organs have
since become too small and weak to support their owners
in the air. Other cases might be cited; for instance,
the Dodo of Mauritius among Pigeons, and the Kakapo
(*Stringops*) of New Zealand among Parrots; but if the
illustrations already given have not convinced you that
disuse of the wings may result in loss of flight, let

me take you finally to the poultry yard, where in the waddling Duck you will see an undeniable instance of degeneration.

As the seat of sexual characters the wing is sometimes most singularly developed or adorned. The males of the Argus Pheasant and Pennant-winged Nightjar have certain feathers enormously lengthened; the Standard-bearer has white plumes growing from the wing; and there are many other cases in which the wing presents sexual characters, not alone through display, but also by use as a musical organ. I do not refer to the whistling sound made by the wings of flying Doves or Ducks, or the humming of Hummingbirds, but to sounds voluntarily produced by birds, and evidently designed to answer the purpose of song.

A simple form of this kind of "music" is shown by the cock in clapping his wings before crowing, in the "drumming" of Grouse, or in the "booming" of Nighthawks, as with wings set they dive from a height earthward. The male Cassique (*Ostinops*) of South America, after giving voice to notes which sound like those produced by chafing trees in a gale, leans far forward, spreads and raises his large orange and black tail, then vigorously claps his wings together over his back, making a noise which so resembles the cracking of branches that one imagines the birds learned this singular performance during a gale.

The birds mentioned thus far have no especial wing structure beyond rather stiffened feathers; but in the Woodcock, some Paradise-birds and Flycatchers, Guans, Pipras, and other tropical birds, certain wing-feathers are singularly modified as musical instruments. Sometimes the outer primaries are so narrowed that little but the shaft or midrib is left, as in both sexes of the Woodcock, when the rapid wing-strokes are accompanied by a

high, whistling sound.   In other cases the shafts of the
wing-feathers may be much enlarged and horny, when
the bird makes a sin-
gular snapping sound
in flight.

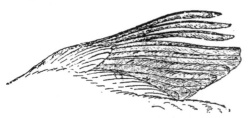

FIG. 9.—Wing of Woodcock, showing three outer attenuate feathers. (¹/₂ natural size.)

If you recall the
supplicating manner of
a young bird as with
gently fluttering wings
it begs for food, you

will recognize one of several ways in which the wings
may express emotion.   Birds also threaten with their
wings, as any hen with chicks will testify, and from this

FIG. 10.—Jacana, showing spur on wing (natural size) and elongated toes (¹/₃ natural size).

gesture to the actual delivery of a blow is but a step.
Swans, Pigeons, and Chickens can deal forcible blows
with their wings.   Screamers, Lapwings, and Jacanas

have formidable spurs on their wings, which they are supposed to use in combat.

The Tail.—Except when sexually developed, the shape of the tail is largely governed by the character of its owner's flight. Male Lyre-birds, Pheasants, Fowls, Hummingbirds, and many others furnish well-marked instances of the tail as a sexual character. Indeed, as the least important to the bird of the four external organs we are speaking of, the tail is more often sexually modified than any of the other three.

The main office of the tail, however, is mechanical, to act as a rudder in flight and a "balancer" when perching. Short-tailed birds generally fly in a straight course, and can not make sharp turns, while long-tailed birds can pursue a most erratic course, with marvelous ease and grace. The Grebes are practically tailless, and their flight is comparatively direct, but the Swallow-tailed Kite, with a tail a foot or more in length, can dash to right or left at the most abrupt angle.

Among tree-creeping birds, which always climb upward, the tail is used as a brace or prop. This character, as has been said, is possessed by all Woodpeckers, by the quite different Woodhewers of South America, the Brown Creepers of temperate regions, and other birds (see Figs. 3 and 4).

The two middle feathers in the tail of the Motmot, of the American tropics, end in a racket-shaped disk, the result of a unique habit. Similarly shaped feathers are found in the tails of some Hummingbirds and Old World Kingfishers, but in the Motmot this peculiar shape is due to a self-inflicted mutilation. The newly grown feathers, as shown in the accompanying figure, lack the terminal disk, but as soon as they are grown, the birds begin to pick at the barbs, and in a short time the shaft is denuded, in some species for the space of an inch, in others for as much as two inches.

This singular habit is practiced by numerous species of Motmots, ranging from Mexico to Brazil. It is therefore of undoubted age, and we can only speculate upon its use and origin. Young birds from the nest, reared

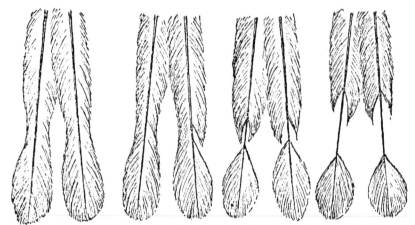

FIG. 11.—Central tail-feathers of Motmot (*Momotus subrufescens*), showing newly grown feathers (at the left) and results of self-inflicted mutilation.

in confinement where they were isolated from others of their kind, trimmed their tail-feathers soon after they were grown.*

The habit, therefore, is inherited, but the mutilation, although it has doubtless been practiced for countless generations, has not become inherent, unless we consider the constriction in the vane of the feather at the place where it is to be trimmed an indication of inheritance.

The Motmot gesticulates with its tail in a remarkable manner, swinging it from side to side, so that it suggests the pendulum of a clock, or sweeping it about in circles with a movement which reminds one of a bandmaster flourishing his baton. We shall find in other species, also, that the tail, more than any other organ, is used to express emotion. Recall its twitching and wagging; how it is nervously spread or " jetted," showing the white

* See Cherrie, The Auk (New York city), vol. ix, 1892, p. 322.

outer feathers, as in the Meadowlark. The tail may also be expressive of disposition. Compare the drooped tail of a pensive Flycatcher with the uptilted member of an inquisitive Wren.

But it is when displaying its beauties that a bird speaks most eloquently with its tail. Can anything exceed the pompous pride of a Turkey cock strutting in swollen glory, with tail stiffly spread? The Peacock erects his tail in a similar manner, but it is entirely concealed by the train of gorgeous feathers which it partially supports.

The Feet.—As the feet share with the wings the responsibilities of locomotion, there is often a close relation between these organs. For example, short-winged terrestrial species like Quails, Grouse, and Rails have well-developed feet, but such aërial creatures as Swifts and Swallows have exceedingly small feet (see Figs. 3 and 4). The aquatic Grebes and Divers are practically helpless on land, but the Ostrich can outrun the horse; while in the perching birds the foot is so specialized that by the auto-

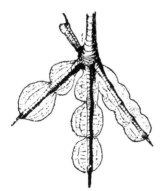

Fig. 12.—Lobed foot of a Coot, a swimming bird of the Rail family. (⅓ natural size.)

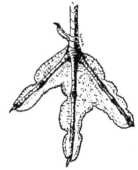

Fig. 13.—Lobed foot of a Phalarope, a swimming bird of the Snipe family. (Natural size.)

matic action of certain tendons the birds are locked to their perches while sleeping. A webbed foot implies ability to swim, and we find this character present in all the water-loving Divers, Auks, Gulls, Cormorants, and Ducks. In the wading Herons and marsh-inhabiting Rails and Gallinules the web is absent, but it reappears in the form of lobes on the toes of the aquatic Coots of the same family.

Some shore-inhabiting Snipe have the bases of the toes united by webs, but the Phalaropes, of two species, have lobed toes not unlike those of the Coots, and are true swimming Snipe living on the sea for long periods.

Length of foot is largely dependent upon length of neck. This is illustrated by the Herons, and is particularly well shown by the long-necked Flamingo, which has a foot twelve inches long. Its toes are webbed, and it can wade in deep water and search for food on the bottom by immersing its long neck and its head.

Fig. 14.—Flamingo, showing relative length of legs and neck in a wading bird. (Much reduced.)

In the tropical Jacanas the toes and toe-nails are much lengthened, enabling the bird to pass over the water on aquatic plants. I have seen these birds walking on small lily leaves, which sank beneath their weight, giving one the impression that they were walking on the water (see Fig. 10).

Many ground-feeding birds use the feet in scratching for food; Chickens are familiar examples. Towhees and

Sparrows use both feet in searching for food, jumping quickly backward and throwing the leaves behind them.

Parrots use their foot as a hand. Some Hawks carry nesting material in it, and all birds of prey strike their quarry with their strongly curved claws, which are then used to carry, or hold it while it is being torn by the bill. The foot of the Fish Hawk is a magnificent organ. The nails are strong and well curved; the inner surface of the toes is set with sharp, horny spikes, and the outer toe is partly reversible, so that the bird grasps its slippery prey from four different points.

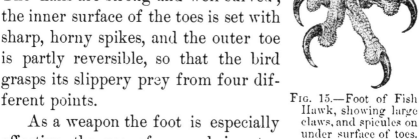

FIG. 15.—Foot of Fish Hawk, showing large claws, and spicules on under surface of toes. (¹/₃ natural size.)

As a weapon the foot is especially effective, the use of spurs being too well known to require comment. Ostriches kick with their feet, and can, it is said, deliver a blow powerful enough to fell a man.

But by far the best instance of modification in the structure of the feet is furnished by Grouse. It is an

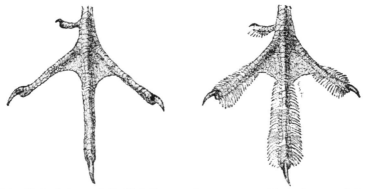

FIG. 16.—Naked toes of Ruffed Grouse in summer; fringed toes of Ruffed Grouse in winter. (²/₃ natural size.)

unusual case of seasonal adaptation in form. During the summer the toes of Grouse are bare and slender, but as these birds are largely ground-haunters, and most of them inhabit regions where the snowfall is heavy, the toes in winter acquire a comblike fringe on either side. Practically, therefore, Grouse don snowshoes in the fall, and wear them until the following spring.

The Bill.—Of the four organs we are considering, the bill is beyond question the most important. We have seen that a bird may be wingless and practically tailless, and may almost lose the use of its feet; but from the moment the bill breaks the eggshell and liberates the chick, the bird's life is dependent upon its services. The variety of offices performed by the bill, and the correspondingly numerous forms it assumes, are, doubtless, without parallel in the animal world.

The special modification of the fore limbs as flight-organs deprives birds of their use for other important services, and consequently we have a biped which, so far as their assistance goes, is without arms or hands. As a result, the duties which would naturally fall to these members are performed by the bill, whose chief office, therefore, is that of a hand.

Occasionally it is sexually adorned, as in the Puffins, several Auks, Ducks, and the White Pelicans, which, during the nesting season, have some special plate, knob, or color on the bill. With the Woodpeckers it is a musical instrument—the drumstick with which they beat a tattoo on some resounding limb. Owls and some other birds, when angry or frightened, snap their mandibles together like castanets. But it is as a hand that the bill gives best evidence of adaptation to or by habit. Among families in which the wings, tail, and feet are essentially alike in form, the bill may present great variation—proof apparently of its response to the demands made upon it.

All birds use it as a comb and brush with which to

perform their toilet, and, pressing a drop of oil from the gland at the root of the tail, they dress their feathers with their bill. Parrots use the bill in climbing, and its hawklike shape in these birds is an unusual instance of similarity in structure accompanying different habits.

Birds which do not strike with their feet may use the bill as a weapon, but the manner in which it is employed corresponds so closely with the method by which a bird secures its food, that as a weapon the bill presents no special modifications. In constructing the nest the bill may be used as a trowel, an auger, a needle, a chisel, and as several other tools.

But as a hand the bill's most important office is that of procuring food; and wonderful indeed are the forms it assumes to supply the appetites of birds who may require a drop of nectar or a tiny insect from the heart of a flower, a snake from the marshes, a clam or mussel from the ocean's beach, or a fish from its waters. The bill, therefore, becomes a forceps, lever, chisel, hook, hammer, awl, probe, spoon, spear, sieve, net, and knife— in short, there is almost no limit to its shape and uses.

With Hummingbirds the shape of the bill is apparently related to the flowers from which the bird most frequently procures its food. It ranges in length from a quarter of an inch in the Small-billed Hummer (*Micro-rhynchus*) to five inches in the Siphon-bill (*Docimastes*), which has a bill longer than its body, and is said to feed from the long-tubed trumpet

Fig. 17.—Decurved bill of Sickle-bill Hummingbird. (Natural size.)

flowers. The Avocet Hummer (*Avocettula*) has a bill curved slightly upward, but in the Sickle-billed Hummer (*Entoxeres*) it is curved downward to form half a circle, and the bird feeds on flowers having a similarly curved corolla. In the Tooth-billed Hummer (*Androdon*) both mandibles are finely serrate at the end, the upper one being also hooked, and the bird feeds on insects which it captures on the surface of leaves and other places.

Among the Woodhewers (*Dendrocolaptidæ*) of South America there is fully as much variability, which reflects

equally variable feeding habits. Some species have short, stout, straight bills, others exceedingly long, slender, curved ones. Mergansers, Gannets, An-hingas, and other birds

FIG. 18.—Serrate bill of Merganser, a fish-eating bird. (½ natural size.)

that catch fish by pursuing them under water, have sharply serrate mandibles, which aid them in holding their slippery prey.

Some shore birds (*Limicolæ*) use the bill as a probe,

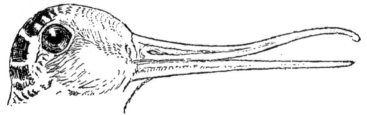

FIG. 19.—Probelike bill of Woodcock, showing extent to which upper mandible can be moved. (⅔ natural size.)

when it may be six inches in length and straight, or curved downward. It has recently been learned that

FIG. 20.—Recurved bill of Avocet. (⅔ natural size.)

several of these probing Snipe, notably the Woodcock, have the power of moving the end of the upper mandible, which better enables them to grasp objects while probing. In the Avocet the bill is curved upward, and the bird swings it from side to side, scraping the bottom in its search for food. The New Zealand Wrybill has its bill turned to the right for the terminal third, and the bird uses it as a crooked probe to push under stones in hunting for its prey. The Siberian Spoonbill Sandpiper has a most singular

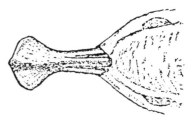

Fig. 21.—Bill of Spoonbill Sandpiper. (Natural size.)

bill, which is much enlarged at the end, suggesting a flat-ended forceps. The Roseate Spoonbill, an entirely different bird, has a somewhat similarly shaped bill, a striking instance of the occurrence of the same form in families which are not closely related.

But probably the most remarkable instance of relation

Fig. 22.—Curved bill of female, straight bill of male Huia-bird. (½ natural size.)

between the form of the bill and feeding habits is furnished by the Huia-bird of New Zealand. The male of this species has a comparatively short, straight bill, while

that of the female is long and curved. The birds feed
on larvæ, which they find in dead wood. The male
hammers and chisels away the wood very much as Woodpeckers do, while the female uses her bill as a probe.
We have, therefore, the singular case of two forms of
the bill arising in the same species as a result of or caus-
ing a corresponding difference in habit.

CHAPTER III.

COLORS OF BIRDS.*

THE almost endless range of variation in the colors and pattern of coloration of birds' plumage has attracted the attention of many philosophic naturalists. Why, for example, should birds from some regions always be darker than those from other regions; why should ground-inhabiting birds generally wear a dull or neutral tinted costume; and why should the male, with few exceptions, be brighter than the female?

For answer I will outline some of the leading facts and theories in connection with this interesting subject. In the first place, however, it will be necessary for us to have some idea of the extent of individual change in color, that is, the various phases of color, which a bird may pass through during different periods of its life.†

---

* Consult Poulton, Colors of Animals (D. Appleton & Co.). Gadow, in Newton's Dictionary of Birds—articles, Color and Feathers. Beddard, Animal Coloration (Macmillan Co.). Keeler, Evolution of the Colors of North American Land Birds; occasional papers, California Academy of Sciences (San Francisco), iii, 1893. Also Allen, reviews of last two works, The Auk (New York city), x, 1893, pp. 189-199, 373-380. Allen, Alleged Changes of Color in the Feathers of Birds without Molting; Bulletin of the American Museum of Natural History, New York city, viii, 1896, pp. 13-44. Chadbourne, Individual Dichromatism in the Screech Owl; The Auk, xiii, 1896, pp. 321-325, and xiv, 1897, pp. 33-39, one plate.

† The term *color*, as here used, means practically the plumage or dress of birds.

Color and Age.—All birds have a special nestling plumage. With those that run or swim at birth, such as Grouse, Snipe, and Ducks, this is a full suit of down, which may be worn for several weeks. With those birds which are helpless when hatched—for instance, Robins, Sparrows, and Orioles—this downy covering is so scanty that they are practically naked. This birth dress is followed by a new growth, known as the "first plumage." Down-covered birds do not acquire this for some time, but with those birds that are born nearly naked it begins to grow soon after they are hatched, and is almost complete when they leave the nest. The first plumage is often unlike that of either parent; for example, the spotted plumage of the Robin. It is worn for several months by some species—certain Snipe and others —but with most land birds it is soon exchanged for the costume they will wear through the winter, usually termed the "immature plumage." This may resemble that of either parent respectively—that is, immature males may be like adult males and immature females like adult females, as with the Bob-white and Cardinal Grosbeak; or the immature birds of both sexes may resemble the adult female, as with the Hummingbird and Bobolink. Again, the immature birds of both sexes may be unlike either of the adults, as with the Eagle and most Hawks; or the immature female may resemble the adult female, while the immature male is unlike either parent, as in the case of the Rose-breasted Grosbeak and Scarlet Tanager. When both parents are alike, the young generally resemble them, and this happens among most of our land birds; for example, the Flycatchers, Crows and Jays, many Sparrows, Vireos, Wrens, and Thrushes.

Immature birds, differing from the adults, may acquire the adult plumage the next spring, as with the Bobolink, or they may then don a second or transition plumage, and not assume the dress of maturity until the second or even the third spring, which is the case with the Orchard Oriole.

Color and Season.—Quite apart from the changes in color due to age, a bird may throughout its life change costumes with the seasons. Thus, the male Bobolink after the nesting season, exchanges his black, white, and buff nuptial suit for a sparrowlike dress resembling that of his mate. The Scarlet Tanager sheds his gay body plumage and puts on the olive-green colors of the female, without changing, however, the color of his black wings and tail. The following spring both birds resume the more conspicuous coats. A more or less similar change takes place among many birds in which the male is brighter than the female, but, among land birds, when the adults of both sexes are alike, there is little or no seasonal change in color.

The Molt.*—These changes in plumage, as far as they are understood, are accomplished by the molt, frequently followed by a wearing off of the differently colored terminal fringe which is found on the new feathers of some birds. It has been stated that birds change color without changing their plumage, either by a chemical alteration in the pigment of the feathers resulting in a new color, or by the actual gain of new pigment from the body; but I know of no instance in which this has been proved, nor do I believe that the latter change is possible. The whole subject offers an excellent field for observation and experiment.

There is a great and as yet but little understood variation in the molting of birds. Not only may closely

---

* See Stone, The Molting of Birds, with Special Reference to the Plumages of the Smaller Land Birds of Eastern North America. Proceedings of the Philadelphia Academy of Natural Science, 1896, pp. 108–167, two plates.

related species molt differently, but the manner and time of molting among individuals of the same species may vary according to their sex, age, and physical condition.

At the close of the nesting season *all* birds renew their entire plumage by molting. The following spring, before the nesting season, most birds molt their body feathers, retaining those of the wing and tail. A few, however, like the Bobolink, have a complete molt at this season also. Others molt only a few of the body feathers, while some birds are adorned at this season with special nuptial plumes.

The beautiful aigrette plumes of the Heron constitute a nuptial dress of this kind. It is for these plumes that the birds have been slaughtered in such enormous numbers that if the demand continues they will speedily become extinct.

Some birds, whose fall plumage is edged with a differently colored tip to each feather, do not molt in the spring, but acquire their wedding dress by the slow wearing off of the fringes to the feathers which have dis-

October.   January.   March.   June.

FIG. 23.—Feathers from back of Snowflake, showing seasonal changes in form and color due to wearing off of tips. (Natural size.)

guised them during the winter. The Snowflake, for instance, changes from brown and brownish white to pure black and white by losing the brown tips which have concealed the black or white bases of his feathers.

Much remains to be learned on this subject of the molt, and, although confinement is known to affect its manner and extent, I believe intelligent observation of caged birds will lead to really valuable results.

Color and Food.—In some instances it is known that a bird's color is affected by the nature of its food. It is a common practice among bird fanciers to alter the color of Canaries from yellow to orange-red by feeding them on red pepper. This food, however, is said to have no effect upon adult birds, but must be fed to nestlings. Sauermann's experiments, as quoted by Beddard, show that the red color is not caused by the capsicin or red pigment in the pepper, but by a fatty substance termed triolein. Fed to white fowls, their breasts became red, while the rest of the plumage remained unchanged. It is also stated that dealers alter the color of green Parrots to yellow by feeding them on the fat of certain fishes.

Flamingoes and Scarlet Ibises when kept in captivity lose their bright red colors and become dingy pink or even soiled white, and some animal dealers have acquired a reputation for restoring their natural tints by supplying them with food the nature of which is kept a secret.

Our Purple Finch turns to yellow in captivity. An adult male now in my possession is undergoing his second molt since capture a year ago, and it will evidently leave him without a single red feather. Other wild birds when caged are known to assume more or less abnormal plumages, due, it is supposed, to change in food. There is, however, very little exact information on this subject, and it offers an excellent opportunity for the patient investigator.

Color and Climate.*—Color is a much more variable character than form. There are but few instances in

---

* Read Allen, Bulletin of Museum of Comparative Zoölogy (Cambridge, Mass.), vol. ii, No. 3, 1871, pp. 186-250.

which we can show the cause of a given structure ; but color responds more quickly to the influence of surroundings, and in many cases we can point to cause and effect with some certainty.

This is best illustrated by the relation between climate and color.   Briefly, it has been found that birds are darkest in humid regions and palest in arid regions.

This at first thought seems of small moment, but in reality it is one of the most important facts established by ornithologists.   It is an undeniable demonstration of "evolution by environment"—that is, the bird's color is in part due to the conditions under which it lives.

For example, our common Song Sparrow, which inhabits the greater part of North America, varies so greatly in color in different parts of its range that no less than eleven subspecies or geographical races are known to ornithologists.   The extremes are found in the arid deserts of Arizona, where the annual rainfall averages eight inches, and on the humid Pacific coast from Washington to Alaska, where the annual rainfall averages about eighty inches.

The Arizona Song Sparrows are pale, sandy colored birds, while those from Alaska are dark, sooty brown. One would imagine them to be different species ; but unlike as are these extremes, they, with the other nine races in this group, are found to intergrade in those regions where the climatic conditions themselves undergo a change.   That is, as we pass from an arid into a humid region, the birds gradually get darker as the average rainfall increases.

If now we study other birds living in these regions, we find that many of them, especially the resident species, are similarly affected by the prevailing climatic influences—that is, many Arizona birds are bleached and faded in appearance, while all the thirty odd Northwest

Pacific coast races are darker or more heavily streaked or barred than any of their congeners. It is of importance to observe that these differences are shown by young birds in fresh plumage—evidence that the characters acquired through climate have been inherited.

There are many similar cases, but some species seem more easily affected than others, and throughout their ranges are markedly affected by the conditions under which they live. Thus we have nine races of Screech Owl, eleven of Horned Lark, six of Junco, etc.

These races, or subspecies, are species in process of formation. The extremes are still connected by intermediate or natural links, but if, through any cause, these intermediates should disappear, the extremes would then be left as distinct species.

Color and Haunt and Habit.—The relation of a bird's color to its haunts and habits is a complex subject. Any attempt at its explanation should be based on so exact a knowledge of the *facts* in the case, that I can not too strongly emphasize here the necessity for observations in the field. Only a close study of the living bird will justify us in advancing theories to account for its coloration.

Many explanations have been offered to account for certain colors and markings of birds, but often, I fear, without adequate knowledge of the bird's habits. I shall speak of only four classes of colors; they are *protective, deceptive, recognition,* and *sexual colors.*

Protective colors render a bird inconspicuous in order that it may escape its enemies. Deceptive colors render it inconspicuous in order that it may more easily approach its prey. In both cases the bird should harmonize in color with its immediate surroundings.

A survey of the birds of the world shows that on the whole this is true. Thus almost all ground-inhabiting birds, such as Snipe, Plover, Quail, Grouse, Sparrows, are generally dull brown or gray, like the ground, leaves, or grasses about them, while tree-haunting birds, especially those that live in the foliage or feed from blossoms, are, as a rule, brightly colored. In this class belong Hummingbirds, Orioles, the gayer-plumaged Finches, Tanagers, Warblers, and many others. It is partly owing to this fact that the erroneous idea concerning the brilliant plumage of all tropical birds has become established. The rich vegetation of the tropics furnishes a home to a far greater number of brightly colored birds than are found in temperate regions; still, they are not more numerous than the dull-colored species that live on the tree trunks, in the undergrowth, or on the ground, where, owing to the nature of both their colors and haunts, they are likely to be overlooked.

Between these two extremes there are numerous intermediate groups, most of which conform to the general law of protective coloration. There are, it is true, exceptions, but every close student of bird-life must be so impressed with the dangers to which birds are exposed, that he can not doubt that the chief object of color is usually for its wearer's concealment.

The term "protective coloration" has lately received fresh significance through the studies of Mr. Abbott H. Thayer.* Mr. Thayer proves conclusively that protective coloration lies not so much in an animal's resemblance in color to its surroundings as in its gradation of color. Thus he points to the fact that, as a rule, animals are darker above than below—that is, those parts receiving the most light are darkest, while the parts receiving the least light are palest. In effect it follows that the darker upper parts are brightened, while the paler under parts are

---

* See his papers on The Law which Underlies Protective Coloration, in The Auk (New York city), vol. xiii, pp. 124-129, 318-320, eleven figures.

darkened, the result being a uniform color, with an apparent absence of shadow, tending to render the object invisible.

Mr. Thayer clearly demonstrates his discovery by using several decoys about the size and shape of a Woodcock's body. These he places about six inches above the ground on wire uprights, or in a row on a horizontal rod. One of these decoys he colors uniformly, above and below, to resemble the earth about it, or he may even give it a fine coating of the earth itself. The upper half of the other decoys is treated in exactly the same manner, but their lower half is graded to a pure white on the median line below. At a distance of forty or fifty yards the uniformly colored decoy can be plainly seen, but those which are white below are entirely invisible until one is within twenty or thirty feet of them.

After definitely locating these graded decoys the experiment may be repeated; but the result will always be the same. As one slowly retreats from them they will, as by magic, seem to pass out of existence, while the one which is colored alike both above and below can be seen distinctly.

One of the best arguments for the value of a protective coloration is the fact that the birds themselves are such thorough believers in it. Here we have the reason why—in sportsman's parlance—game birds "lie to a dog." When there is sufficient cover, they trust to their protective coloring to escape detection, and take wing only as a last resort; but when cover is scanty, they generally rise far out of gunshot. Some Snipe and Sparrows, however, attempt to conceal themselves even on bare sand or worn grass by squatting close to the earth, with which their plumage harmonizes in color.

A sitting Woodcock had such confidence in its own invisibility that it permitted itself to be stroked without leaving the nest; but when a light snow fell, and the bird became a conspicuous dark object against a white background, it took wing on the first suspicion of danger.

I could mention many other similar instances, but the careful observer will soon find them included in his own experience.

Deceptive, or, as Poulton terms it, "aggressive" coloration is perhaps best illustrated by common Flycatchers (*Tyrannidæ*). Although these birds live in and about trees, they are, as a rule, quietly attired in olive-green or olive-gray, and are quite unlike the brilliantly clad, *fruit-* eating Tanagers, Orioles, Parrots, and other birds that may be found near them. Insects are therefore more likely to come within snapping distance than if these birds were conspicuously colored. In the same manner we may explain the colors of Hawks, which are never brightly plumaged.

It is well known that many arctic animals become white on the approach of winter. With Ptarmigans this is doubtless an instance of protective coloration, but the Snowy Owl, who feeds on the Ptarmigan, may be said to illustrate deceptive coloration.

Recognition, signaling, or directive colors have, with more or less reason, been made to include many different types of markings, of which I shall mention only those that are conspicuously shown in flight or by some movement. Such are the white outer tail-feathers of Juncos, Meadowlarks, Towhees, and many other birds, and certain wing and rump patches, which are noticeable only when the bird is on the wing. Markings of this kind are supposed to aid birds in recognizing others of their kind, their special use being to keep the individuals of a family or flock together, so that when one starts the others can readily follow. The theory is open to objections, but these so-called recognition marks are so often found among birds that they doubtless are of some use, though their exact value remains to be determined by closer observation.

Color and Sex.*—It is not possible here to discuss at length the vexed question of sexual coloration. But, as a means of directing observation, I present a synopsis of the principal types of secondary sexual characters, with some of the theories which have been advanced to account for them.

SYNOPSIS OF THE SECONDARY SEXUAL CHARACTERS OF BIRDS.

I. STRUCTURAL.

Size.
{ Male larger than female (usual).
{ Female larger than male (rare).

Plumage.
{
  Color. { Male brighter than female.
         { Female brighter than male (rare).
  Form. { Assumption of plumes, ruffs, crests, trains, etc.: special modification of wing and tail feathers.
         { a. Worn by male alone.
         { b. Worn by both sexes.
}

Of the body. { Sole or greater development in male of brightly colored bare tracts of skin, combs, wattles, caruncles, and other fleshy or horny appendages.

Of the feet. Sole or greater development in male of spurs.

Of the bill. { Male with more highly colored or larger bill than female.

II. FUNCTIONAL.

Pursuit.
{ By male when similar to or brighter than female.
{ By female when brighter than male.

Display. By male of accessory plumes and other appendages.

Battle. By male using spurs, wings, bill, etc.

Music.
{ Vocal, by male and, rarely, female.
{ Mechanical, by male and sometimes female.

Special habits.
{ Dances, mock fights, aërial evolutions, construction of bowers, decoration of playgrounds, attitudinizing, strutting, etc.
{ a. By male before the female.
{ b. Among the males alone.

---

* Read Darwin, The Descent of Man and Selection in Relation to Sex (D. Appleton & Co.). Wallace, Darwinism (Macmillan Co.).

In explanation of these remarkable differences of form and habit, we have first Darwin's theory of "sexual selection." This is based upon the ardor in love, the courage and rivalry of the males, and also upon the powers of perception, taste, and will of the female.

The spurs of the male, for example, are supposed to have been developed through the battles of the males. At first a mere knob, they were an advantage to the bird possessing them, enabling him to defeat his rivals. The successful male would be more likely to have offspring who would inherit the tendency of their father to have spurs, and thus, by selection, the unspurred cocks would gradually be replaced by those better armed. This is known as the "law of battle."

But the bright colors and gay plumes of the cock have originated, under this theory, through the taste of the female, who, it is assumed, would be more likely to accept the attentions of a bird pleasing in her eye than one who was less strikingly adorned. This has been termed by Lloyd Morgan "preferential mating."

Wallace has accepted the law of battle as an effective agent in producing certain characters, but considers it *natural*, rather than *sexual* selection, and he denies the existence of any important evidence proving female selection. He therefore attributes many secondary sexual characters to a surplus of vital energy, which, because of a bird's perfect adaptation to the conditions of its existence, can expend itself in the production of bright colors and ornamental plumes without injury to their owners. That is to say, Wallace ascribes to the action of natural selection any secondary sexual character which is of practical use to the male in conflicts with a rival, but denies the female any part in the matter of pairing. Darwin, as I have said, attributes to the female an æsthetic taste which renders the brilliant colors or display of the male an attractive sight, influencing her choice of a mate.

There is thus a practical agreement in the views of these naturalists on the origin of those sexual characters which may be classed as weapons, and this opinion is, I believe, generally accepted. But the question of female preference, and its influence on the development of bright colors and accessory plumes, still lacks confirmation. Here is an opportunity for every one who can watch wild birds mating.

CHAPTER IV.

THE MIGRATION OF BIRDS.*

To the field student the season of migration is the most interesting of the year. The bird-life of a vast area then passes in review before him. Though living in a temperate region, he may see birds whose summer home is within the Arctic Circle, whose winter haunts are in the tropics. Who can tell what bird he may find in the woods he has been exploring for years?

The comparative regularity with which birds come and go gives an added charm to the study of migration. Their journey is not a "helter-skelter" rushing onward, but is like the well-governed march of an army. We feel a sense of satisfaction in knowing when we may expect to greet a given species, and a secret elation if we succeed in detecting it several days in advance of other observers. We study weather charts, and try to foretell or explain those great flights or "waves" of birds which are so closely dependent upon meteorologic conditions.

---

* Read Allen, Scribner's Magazine, vol. xxii, 1881, pp. 932–938, Bulletin of Nuttall Ornithological Club (Cambridge, Mass.), vol. v, 1880, pp. 151–154. Scott, ibid., vol. vi, 1880, pp. 97–100, Brewster, Memoirs of Nuttall Ornithological Club, No. 1, pp. 22. Cooke and Merriam, Bird Migration in the Mississippi Valley (Washington, 1888). Chapman, The Auk (New York city), vol. v, 1888, pp. 37–39 ; vol. xi, 1894, pp. 12–17. Loomis, ibid., vol. ix, 1892, pp. 28–39 : vol. xi, 1894, pp. 26–39, 94–117. Stone, Birds of Eastern Pennsylvania and New Jersey, pp. 15–28.

Extent of Migration.—The extent of a bird's migration is, in most cases, dependent upon the nature of its food. Birds that are resident in one place throughout the year generally change their fare with the season, and apparently feed with equal relish on seeds or insects. Those that are dependent upon fruit must migrate far enough to find a supply of berries, while the insect-eaters are obliged to travel even farther south.

Most of the migratory birds of our Western States pass the winter in Mexico. Our Eastern Sparrows and our berry-eaters, like the Robin and Bluebird, winter from the Middle States to the Gulf coast, while the majority of our purely insectivorous species cross to Cuba and winter in the West Indies, or continue to Central America and even northern South America. Snipe and Plover make the most extended migrations, some species breeding within the Arctic Circle and wintering along the coasts of Patagonia.

Times of Migration.—Let us suppose we are about to observe the spring migration of birds at Englewood, New Jersey—a few miles from New York city. Birds arrive here about a week later than at Washington, D. C., and a week earlier than at Boston.

During January and February, while watching for some rare visitor from the North, we shall find that Tree Sparrows and Juncos are everywhere common. Less frequently we may see Shrikes, Winter Wrens, Golden-crowned Kinglets and Brown Creepers, and rarely Snowflakes, Red Crossbills, and Redpolls will be observed. These birds are winter visitants, coming to us from the North in the fall and leaving in March and April.

Of course, in addition to these migratory birds, we shall see most if not all of our commoner permanent residents, or the birds which are with us throughout the year. They are the Bob-white, Ruffed Grouse, Red-shouldered and Red-tailed Hawks, Barred and Screech Owls, Downy and Hairy Woodpeckers, Blue Jay, Crow, Goldfinch, Song Sparrow, White-breasted Nuthatch, and Chickadee.

Generally speaking, the birds in the front rank of the feathered army which soon will invade the land are those whose winter quarters are farthest north, while those that winter farthest south bring up the rear.

From February 20 to March 10, therefore, we may expect to see Purple Grackles, Robins, Bluebirds, and Red-winged Blackbirds; birds that have wintered but a short distance south of us—if not with us—and who have accepted the slightest encouragement from the weather as an order to advance. All the first comers will doubtless be males, this sex, as a rule, preceding the females by several days.

About the middle of March we may look for the Woodcock, Meadowlark, Fox Sparrow, Cowbird, and Phœbe; their time of arrival being largely dependent upon the temperature—warm weather hastening, and cold weather retarding their movements.

Toward the last of March, Wilson's Snipe, the King-fisher, Mourning Dove, Swamp and Field Sparrows are due.

Early in April the Purple Finch, White-throated, Vesper, and Chipping Sparrows will announce their return in familiar notes, and at the same time Tree Swallows, Myrtle Warblers, Pipits, and Hermit Thrushes will appear. They will soon be followed by Barn Swallows and Ruby-crowned Kinglets.

The migration is now well under way, and we go to the field with the assurance of meeting some lately arrived feathered friend almost daily. Between April 20 and 30 we will doubtless note among the newcomers, the Green Heron, Spotted Sandpiper, Whip-poor-will,

Chimney Swift, Least Flycatcher, Towhee, Purple Martin, Cliff and Bank Swallows, Black and White and Black-throated Green Warblers, Oven-bird, House Wren, Brown Thrasher, Catbird, and Wood Thrush. This troop surely is not without musicians. In ringing tones they herald the victory of Spring over Winter.

The season of cold waves has passed, and the birds now appear with the regularity of calendar events. From May 1 to 12 the migration reaches its height. It is a time of intense interest to the bird student, and happy is he who can spend unlimited time afield. Some mornings we may find ten or more different species that have come back to us, and each one may be represented by many individuals. The woods are thronged with migrants, and the scantily leaved trees and bushes enable us to observe them far more easily than we can when they travel southward in the fall. During this exciting period we should see the Cuckoos, Nighthawk, Ruby-throated Hummingbird, Crested Flycatcher, Kingbird, Wood Pewee, Baltimore and Orchard Orioles, Bobolink, Indigo Bunting, Rose-breasted Grosbeak, Scarlet Tanager, Red-eyed, Warbling, Yellowthroated, and White-eyed Vireos, Long-billed Marsh Wren, Wilson's Thrush, Redstart, Yellow-breasted Chat, Maryland Yellow-throat, Yellow Warbler, and others of its family.

Succeeding days will bring additions to the ranks of these species, and there will also be numerous small Warblers to look for, but by May 12 all our more familiar and common birds have arrived. During the rest of the month, as the transient visitants, or species which nest farther north, pass onward, birds gradually decrease in numbers, and by June 5 we have left only those that will spend the summer with us.

The migration over, we can now give our whole

attention to a study of nesting habits. As a matter of fact, the nesting season begins quite as early as the spring migration, the Great Horned Owl laying its eggs late in February. In March and April other birds of prey and the earlier migrants nest. May migrants go to housekeeping soon after they reach their old homes, and by June 5 there are few species that have not nests.

With birds that rear two or three broods, the nesting season may extend into August. With those that have but one brood it may be over early in July. At this time we begin to miss the jolly, rollicking music of the Bobolink. Soon he will leave the meadow he has animated for two months, and with his young join growing flocks of his kind in the wild-rice marshes. His handsome suit of black and white and buff will be exchanged for the sparrowlike Reedbird dress, and in place of the merry song he will utter only a metallic *tink*. This note is characteristic of the season. Day and night we hear it from birds high in the air as they hasten to their rendezvous in the marshes.

July 1, Tree Swallows, who nest rarely if at all near New York city, appear and establish their headquarters in the Hackensack meadows—a first step on the migratory journey. July is a month for wanderers. The nesting season of most one-brooded birds is over; they are not yet ready to migrate, and pass the time roving about the country with their families.

In August birds are molting and moping. The careful observer will find that a few Warblers and Flycatchers have returned from the north and are passing southward, but, as a rule, August is a month to test the patience of the most enthusiastic bird student. Late in the month migrants become more numerous, but between the " flights " or " waves " there are days when one may tramp the woods for miles without seeing a dozen birds.

September is the month of Warblers. They come in myriads during the latter half of the month, and on favorable nights we may sometimes hear their fine-voiced *tseeps* as they fly by overhead. About the 25th of the month our winter residents, the Junco, Winter Wren, Golden Kinglet, and Brown Creeper, will arrive.

The summer residents are now rapidly leaving us. In a general way it may be said that the last birds to arrive in the spring are the first to leave in the fall, while the earliest spring migrants remain the longest.

October and November are the months of Sparrows. They rise in loose flocks from every stubble or weed field, and seek shelter in the bordering bushy growth. Should the season prove warm, many of these hardy seed-eaters will stay with us well into December, but at the first really cold weather they retreat southward.

This completes the merest outline of the movements of our migratory birds. It will be seen that in reality there are but few periods during the year when some event is not occurring in the bird world. As we accumulate records for comparison, and learn to appreciate their meaning, our interest in the study of migration will increase and be renewed with the changing seasons.

We have found, in this brief review, that our birds may be placed in four classes, as follows:

1. *Permanent Residents.*—Birds that are represented in the same locality throughout the year.

2. *Summer Residents.*—Birds that come to us in the spring, rear their young, and depart in the fall.

3. *Winter Residents.*—Birds that come from the north in the fall, pass the winter with us, and return to their more northern homes in the spring.

4. *Transient Visitants.*—Birds whose summer home is north and whose winter home is south of us. In traveling from one to the other they pass through the intervening region as " transients."

Manner of Migration.—The Oriole, who builds his swinging nest in your elm tree, will winter in Central America; the Bobolink, who seems so care-free in your meadows, must journey to his winter quarters in southern Brazil. But, unless accident befalls, both birds will return to you the following spring. We are so accustomed to these phenomena that we accept them as part of the changing seasons without realizing how wonderful they are. But look for a moment at a map, and try to form a mental picture of the Bobolink's route. Over valleys, mountains, marshes, plains, and forests, over straits and seas hundreds of miles in width, he pursues a course through trackless space with a regularity and certainty which brings him to the same place at nearly the same time year after year. How much of his knowledge of the route he has inherited, and how much learned during his own lifetime, is a question we may return to later; now we are concerned with actual methods of migration.

Immediately after, or even during the nesting season, many birds begin to resort nightly to roosts frequented sometimes by immense numbers of their kinds, with often the addition of other species. These movements are apparently inaugurated by the old birds, and are in a sense the beginnings of the real migratory journey. Other birds roam the woods in loose bands or families, their wanderings being largely controlled by the supply of food.

During this time they may be molting, but when their new plumage is acquired they are ready for the start. The old birds lead the way, either alone or associated with the young. Some fly by day, some by night, and others by both day and night. This fact was first established by Mr. William Brewster, who, in his admirable memoir on Bird Migration, writes: " Timid, sedentary, or feeble-winged birds migrate by night, because they are either afraid to venture on long, exposed journeys by daylight, or unable to continue these journeys day after day without losing much time in stopping to search for food. By taking the nights for traveling they can devote the days entirely to feeding and resting in their favorite haunts. Good examples are Thrushes (except the Robin), Wrens, Warblers, and Vireos.

" Bold, restless, strong-winged birds migrate chiefly, or very freely, by day, because, being accustomed to seek their food in open situations, they are indifferent to concealment, and being further able to accomplish long distances rapidly and with slight fatigue, they can ordinarily spare sufficient time by the way for brief stops in places where food is abundant and easily obtained. Under certain conditions, however, as when crossing large bodies of water or regions scantily supplied with food, they are sometimes obliged to travel partly, or perhaps even exclusively, by night. Excellent examples are the Robin (*Merula*), Horned Lark (*Otocoris*), and most *Icteridæ* [Bobolink, Blackbirds, and Orioles].

" Birds of easy, tireless wing, which habitually feed in the air or over very extensive areas, migrate exclusively by day, because, being able either to obtain their usual supply of food as they fly, or to accomplish the longest journeys so rapidly that they do not require to feed on the way, they are under no necessity of changing their usual habits. The best examples are Swallows, Swifts, and Hawks."

While migrating, birds follow mountain chains, coastlines, and particularly river valleys, all of which become

highways of migration.   Through telescopic observations
it has been learned that migrating birds travel at a great
height.   The exact height remains to be determined, but
it is known that many migrants are at least a mile above
the earth.   From this elevation they command an extended view, and in clear weather prominent features of
the landscape are doubtless distinguishable to their powerful vision at a great distance.

It is when fogs and storms obscure the view that birds
lose their way.   Then they fly much lower, perhaps seeking some landmark, and, should a lighthouse lie in their
path, they are often attracted to it in countless numbers.
Thousands of birds perish annually by striking these
lights during stormy fall weather.   In the spring the
weather is more settled and fewer birds are killed.

Although birds are guided mainly by sight, hearing is
also of assistance to them on their migrations.   Indeed, at
night, young birds, who have never made the journey before, must rely largely upon this sense to direct them.   It
is difficult for us to realize that on favorable nights during
the migratory season myriads of birds are passing through
the dark and apparently deserted air above us.   Often
they are so numerous as to form a continuous stream, and
if we listen we may hear their voices as they call to one
another while flying rapidly onward.

Some idea may be formed of the multitude of birds
which throng the upper air on favorable nights during
their migration by using a telescope.   One having a two-
inch object glass will answer the purpose.   It should be
focused on the moon, when the birds in passing are sil-
houetted against the glowing background.   At the proper
focal distance they appear with startling distinctness.   In
some cases each wing-beat can be detected, and with a
large glass it is even possible to occasionally recognize
the kind of bird.

Observations of this kind should be made in September, when the fall migration is at its height. On the night of September 3, 1887, at Tenafly, New Jersey, a friend and myself, using a six-and-a-half-inch equatorial glass, saw no less than two hundred and sixty-two birds cross the narrow angle subtended by the limbs of the moon between the hours of eight and eleven. Observations made several years later, in September, from the observatory of Columbia University, yielded closely similar results.

This nocturnal journey of birds may also be studied from lighthouses. On September 26, 1891, I visited the Bartholdi Statue of the Goddess of Liberty, in New York Bay, for this purpose. The weather was most favorable. The first bird was observed at eight o'clock, and for the succeeding two hours others were constantly heard, though comparatively few were seen. At ten o'clock it began to rain; and almost simultaneously there was a marked increase in the number of birds about the light, and within a few minutes there were hundreds where before there was one, while the air was filled with the calls of the passing host.

From the balcony which encircles the torch the scene was impressive beyond description. We seemed to have torn aside the veil which shrouds the mysteries of the night, and with the searching light exposed the secrets of Nature.

By far the larger number of birds hurried onward; others hovered before us, like Hummingbirds before a flower, then flew swiftly by into the darkness; and some, apparently blinded by the brilliant rays, struck the statue slightly, or with sufficient force to cause them to fall dead or dying. At daybreak a few stragglers were still winging their way southward, but before the sun rose the flight was over.

Origin of Migration.—Why do birds migrate? It is true that in temperate and boreal regions the return of cold weather robs them of their food, and they retreat southward. But many, in fact most, birds begin their southern journey long before the first fall frost. We have seen that some species start as early as July and August. Furthermore, there are many birds that come to our Gulf and South Atlantic States to nest, and when the breeding season is over they return to the tropics. Surely, a lower temperature can not be said to compel them to migrate. Even more remarkable than the southward journey in the fall is the northward journey in the spring. Our birds leave their winter homes in the tropics in the height of the tropical spring, when insect and vegetable food is daily increasing. They leave this land of plenty for one from which the snows of winter have barely disappeared, often coming so early that unseasonable weather forces them to retreat.

I believe that the origin of this great pilgrimage of countless millions of birds is to be found in the existence of an annual nesting season. In my opinion, it is exactly paralleled by the migration of shad, salmon, and other fishes to their spawning grounds, and the regular return of seals to their breeding rookeries.

Most animals have an instinctive desire for seclusion during the period of reproduction, and when this season approaches will seek some retired part of their haunts or range in which to bring forth their young. Salmon may travel a thousand miles or more from the ocean, and, leaping the rapids or other barriers in their way, finally reach the headwaters of some river where their eggs may be deposited in safety. Seals migrate with regularity to certain islands, where their young are born. Even our domesticated Hens, Turkeys, Ducks, and Peafowl, if given freedom, will travel a greater or less distance in search of a place where they may conceal their nests.

Many species of tropical sea birds resort each year to some rocky islet, situated perhaps in the heart of their range, where they may nest in safety. This is not migration as we understand the word; but, nevertheless, the object is the same as that which prompts a Plover to travel to the arctic regions; moreover, the movement is just as regular. These sea birds pass their lives in the tropics, their presence or absence in any part of their range being largely dependent upon the supply of food. But, as in the case of the Warbler which migrates from South America to Labrador, they are annually affected by an impulse which urges them to hasten to a certain place. This impulse is periodic, and in a sense is common to all birds. There is a regular nesting season in the tropics, just as there is a regular nesting season in the arctic regions.

There is good reason, therefore, for the belief that the necessity of securing a home in which their young could be reared was, as it still is, the cause of migration. It must be remembered, however, that birds have been migrating for ages, and that the present conditions are the result of numerous and important climatic changes. Chief among these is doubtless the Glacial period. Indeed, Dr. Allen has stated, and the theory has been generally accepted, that the migration of birds was the outcome of the Glacial period. That their journeys were greatly increased and the habit of migration extended during the ice age is apparently undeniable, but, although previous to the Glacial period a warm temperate climate prevailed nearly to the northern pole, there must even then have been sufficient difference between winter and summer climate to render a certain amount of migration necessary. Furthermore, there is a well-defined migra-

tion in the southern hemisphere, where no evidences of
glaciation have as yet been discovered.

As I have said, the existing conditions are the result
of changes which have been active for ages. No species,
therefore, has acquired its present summer range at one
step, but by gradually adding new territory to its breeding ground. For example, certain of our Eastern birds
are evidently derived through Mexico, and in returning
to their winter quarters in Central America, they travel
through Texas and Mexico and are unknown in Florida
and the West Indies. Others have come to us through
Florida, and in returning to their winter quarters do not
pass through either Texas or Mexico. This is best illustrated by the Bobolink, an Eastern bird which, breeding
from New Jersey northward to Nova Scotia, has spread
westward until it has reached Utah and northern Montana. But—and here is the interesting point—these birds
of the far West do not follow their neighbors and migrate
southward through the Great Basin into Mexico, but,
true to their inherited habit, retrace their steps, and leave
the United States by the roundabout way of Florida,
crossing thence to Cuba, Jamaica, and Yucatan, and wintering south of the Amazon. The Bobolinks of Utah
did not learn this route in one generation; they inherited
the experience of countless generations, slowly acquired
as the species extended its range westward, and in returning across the continent they give us an excellent illustration of the stability of routes of migration.

They furnish, too, an instance of one of the most
important factors in migration—that is, the certainty
with which a bird returns to the region of its birth.
This is further evidenced by certain sea birds which
nest on isolated islets to which they regularly return
each year.

Of this wonderful " homing instinct," which plays so

vital a part in the migration of birds, I have no explanation to offer.   We know, however, that it exists not only
in birds but in many other animals.   It is this instinct,
aided by the "heredity of habit," which guides a bird
to its nesting ground.   The Carrier Pigeon is taught its
lines of flight by gradually extending its journeys; a
species establishes its routes of migration by gradually
extending its range.

CHAPTER V.

THE VOICE OF BIRDS.*

Aside from the pleasure to be derived from the calls and songs of birds, their notes are of interest to us as their medium of expression. No one who has closely studied birds will doubt that they have a language, limited though its vocabulary may be.

Song.—Song is a secondary sexual character, generally restricted to the male. With it he woos his mate and gives voice to the joyousness of nesting time. In some instances vocal music may be replaced by instrumental, as in the case of the drumming wing-beat of the Grouse, or the bill-tattoo of the Woodpeckers, both of which are analogous to song.

The season of song corresponds more or less closely with the mating season, though some species begin to sing long before their courting days are near. Others may sing to some extent throughout the year, but the real song period is in the spring.

Many birds have a second song period immediately after the completion of their postbreeding molt, but it usually lasts only for a few days, and is in no sense comparable to the true season of song. This is heralded by the Song Sparrow, whose sweet chant, late in February,

---

* See Witchell, The Evolution of Bird Song (Macmillan Co.). Bicknell, A Study of the Singing of Our Birds; The Auk (New York city), vol. i, 1884, pp. 60-71, 126-140, 209-218, 322-332; vol. ii, 1885, pp. 144-154, 249-262.

is a most welcome promise of spring. Then follow the Robins, Blackbirds, and other migrants, until, late in May, the great springtime chorus is at its height.

The Bobolink is the first bird to desert the choir. We do not often hear him after July 5. Soon he is followed by the Veery, and each day now shows some fresh vacancy in the ranks of the feathered singers, until by August 5 we have left only the Wood Pewee, Indigo Bunting, and Red-eyed Vireo—tireless songsters who fear neither midsummer nor midday heat.

Call-Notes.—The call-notes of birds are even more worthy of our attention than are their songs. Song is the outburst of a special emotion; call-notes form the language of every day. Many of us are familiar with birds' songs, but who knows their every call-note and who can tell us what each call means? For they have a meaning that close observation often makes intelligible.

Listen to the calls of the Robin and learn how unmistakably he expresses suspicion, alarm, or extreme fear; how he signals cheerfully to his companions or gives the word to take wing. Study the calls of the Crow or Blue Jay, and you will find that they have an apparently exhaustless vocabulary.

It is supposed that birds, like men, do not inherit their language, but acquire it. Thus there are recorded instances of young birds who had been isolated from others of their kind, learning to sing whatever song they heard. On the other hand, it is said that a bird inherits its own notes, at least to some extent, and, while it may not sing the song of its species perfectly, its song will still be sufficiently characteristic to be recognizable. There are, however, very few satisfactory observations on this subject, and keepers of cage-birds have here an excellent opportunity for original investigation.

CHAPTER VI.

THE NESTING SEASON.*

IF you would really know birds, you must study them during nesting time. At this season they develop habits that you will be surprised to learn they possess. The humble owner of some insignificant call-note now fills the rôle of a skilled musician. The graceful, leisurely Marsh Hawk gives vent to his feelings in a series of aërial somersaults over the meadows; the sedate, dignified Woodcock tries to express his emotions by means of spiral evolutions which carry him far above his usual haunts; the Night-Hawk dives earthward with needless recklessness; in fact, birds seem inspired by the joy of the season, and all the brightness of a May morning is reflected in their voices and actions.

Mating over, there follow the marvels of nest-building with its combined evidences of instinct and intelligence. In due time the young appear, and the bird, now a parent, abandons the gay habits of the suitor, and devotes every waking moment to the care of its young.

Time of Nesting.—With most birds the nesting season is periodic and annual. With migratory birds it coincides with the season of the year when their summer homes are habitable. But we might suppose that the

---

* Read In Nesting Time, Little Brothers of the Air, and other works by Olive Thorne Miller. A-Birding on a Broncho, by Florence A. Merriam (Houghton, Mifflin & Co.).

permanent residents of the tropics, where seasonal changes are less marked, could nest at any time. Nevertheless, the breeding season in the tropics is as well defined as it is in more northern regions, and occurs with the return of summer, or the season of rains. It is therefore at a time of the year when food is most abundant.

There is an obvious necessity for this regularity. Old birds can wander over large areas in search of food, but the young of many species must be fed in the nest, and their food supply should be both exhaustless and convenient of access.

Among our birds, the Hawks and Owls, whose young are fed on animal food, are the first birds to nest, while those which feed their young on fruit or insects wait until later in the year.

Mating.—Birds are ardent lovers. In their effort to win a bride the males display their charms of song and plumage to the utmost, and will even enter the lists to do battle for the possession of a mate.

It is not possible to describe here the many peculiar customs of birds during the season of courtship. It may simply be said that every bird will then repay the closest observation. For the scientific-minded there is opportunity to secure evidence bearing upon the theory of Natural Selection; for every one there is endless entertainment in the human traits which birds exhibit.

The Nest.—The first step in nest-building is the selection of a site. There is almost no suitable location, from a hole in the ground to branches in the tree-tops, in which birds may not place their nests. Protection seems to be the chief *desideratum*, and this is generally secured through concealment. Most birds hide their nests. Many sea birds, however, lay their eggs on the shores or cliffs, with no attempt at concealment; but, as a rule, birds that nest in this manner resort to uninhabited islets and secure protection through isolation.

Some birds nest alone, and jealously guard the vicinity of their home from the approach of other birds, generally of the same species. Others nest in colonies brought together by temperament or community of interests, and dwell on terms of the closest sociability.

The material used by birds in building their nests is as varied as the nature of the sites they select. The vegetable kingdom contributes much the largest share. Grasses, twigs, and rootlets are the standard materials; but plant-down, plant-fibers, bark, leaves, lichens, clay, spiders' webs, hair, fur, and feathers are also used, while in some cases a gummy secretion of the salivary glands furnishes a kind of glue.

Birds have been classified, according to the manner in which they employ these articles, as weavers, tailors, masons, molders, carpenters, felters, etc.

Sometimes both sexes assist in the construction of the nest, or one bird collects the material while the other adjusts it. Again, the female performs the task alone, aided only by the encouraging voice of the male.

The time of construction varies from one to two weeks to as long as three months in the case of the South American Ovenbird, who in June begins to build the nest it will not occupy until October. The Fish Hawk evidently believes in the value of a stick in time, and often repairs its nest in the fall.

Lack of space prohibits a discussion of the influences which assist in determining the character of birds' nests. They may be summarized as follows:

First, necessity for protection.

Second, conditions imposed by locality. These affect both the site and material, as illustrated by Doves, who nest in trees in wooded countries and on the ground in treeless regions, and by birds who substitute strings, cotton, or rags for their usual nesting materials.

Third, condition of the young at birth, whether feathered or naked. The young of what are termed " præcocial" birds are hatched with a covering of downy feathers. Gulls, Ducks, Snipe, Chickens, Partridges, and Quails are familiar members of this group. Their young can run about soon after birth, and a well-formed nest is not needed. But the young of " altricial" birds are hatched practically naked and are reared in the nest, which is therefore not only a receptacle for the eggs during incubation, but a home. Thrushes, Sparrows, in fact all Perching Birds, Woodpeckers, Hummingbirds, and many others belong in this group of altricial birds.

Fourth, temperament, whether solitary or social. Hawks, fierce and gloomy, nest alone, while the cheery, happy Swallows nest in colonies.

Fifth, structure of the bird. The tools—that is, the bills and feet—of some birds are more serviceable than those of others. We should not expect a Dove to build the woven nest of an Oriole, nor a Hummingbird to fashion a Woodpecker's dwelling.

Sixth, feeding habit. In some few cases feeding habit may determine the character of the nest. For instance, Woodpeckers, in securing their food from trees, often make large excavations, which it is quite natural they should have learned to use as nests.

Seventh, inherited habit, or instinct. There seems no reason to doubt that birds inherit their knowledge of nest-building, for in several cases where birds have been taken from the nest and reared alone, they have afterward constructed a nest resembling that of their species. It would therefore appear that inherited habit is a fact. Through it we may explain not only the similarity in the nests of the same species, but also certain habits for which we can give no satisfactory reason. Thus the Crested Flycatcher's strange custom of using a cast snake-skin in its nesting materials probably originated with the birds in the tropics, where it is still followed by nearly related species of Crested Flycatchers. With them there may-be a reason for this habit, but with our bird, living as it does under entirely different conditions, it is doubtless only an inheritance, surviving even when the necessity for it has ceased to exist.

Eighth, change of habit. Some birds are influenced by changes in their surroundings, and alter their nesting habits when it proves to their advantage to do so. Chimney Swifts, who have exchanged hollow trees, in which they were exposed to their natural enemies, for the comparative safety of chimneys, are good examples. But a far better one is given by that prodigy in feathers, the House Sparrow. Is there any available site in which this thoroughly up-to-date bird will not place its nest? It has taken possession of even the hollow spaces about certain kinds of electric lamps, and has been observed repairing its nest at night by their light!

The Eggs.—Usually, little time is lost between the completion of the nest and the laying of the eggs. The number of eggs composing what oölogists term a full set or clutch ranges from one to as many as twenty. At the time of laying, the ovary contains a large number of partly formed eggs, of which, normally, only the required number will become fully developed. But if the nest be robbed, the stolen egg will frequently be replaced. The long-continued laying of our domestic fowls is an instance of this unnatural stimulation of the ovary. Doubtless the most remarkable recorded case of egg-laying by a wild bird is that of a High-hole or Flicker, who, on being regularly robbed, laid seventy-one eggs in seventy-three days!

The eggshell is composed largely of carbonate of lime,

which is deposited in layers. The final layer varies greatly in appearance, and may be a rough, chalky deposit, as in Cormorants and others, or thin and highly polished, as in Woodpeckers.

The colors of eggs are due to pigments, resembling bile pigments, deposited by ducts while the egg is in the oviduct. One or more of the layers of shell may be pigmented, and variations in the tints of the same pigment may be caused by an added layer of carbonate of lime, producing the so-called " clouded " or " shell markings."

While the eggs of the same species more or less closely resemble one another, there is often so great a range of variation in color that, unless seen with the

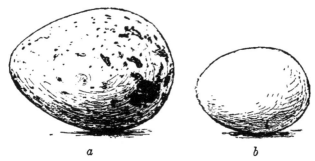

                            a   b

Fig. 24.—Egg of (a) Spotted Sandpiper, (b) Catbird, to show difference in size of eggs of præcocial and altricial birds of same size. (Natural size.)

parent, it is frequently impossible to identify eggs with certainty. The eggs of præcocial birds, whose young are born with a covering of down and can run or swim at birth, are, as a rule, proportionately larger than the eggs of altricial birds, whose young are born in a much less advanced condition. This is illustrated by the accompanying figure of the eggs of the Spotted Sandpiper and the Catbird.

The period of incubation is apparently closely dependent upon the size of the egg, and varies from ten days in the Hummingbird to forty odd in the Ostrich and, it is said, some fifty in the Emu.

Among some species both sexes share equally the task of incubation. In others, the female is longer on the nest, the male taking her place during a short period each day while she is feeding. Less frequently the female is not at all assisted by her mate, and in some cases—Ostriches, Emus, Phalaropes, and a few others—the male alone incubates.

The Young.—The care of the young and their mental and physical development afford us unequaled opportunities for the study of bird character. We may now become acquainted not only with the species but with individual birds, and at a time when the greatest demands are made upon their intelligence.

We may see the seed-eaters gathering insects and perhaps beating them into a pulp before giving them to their nestlings: or we may learn how the Doves, High-holes, and Hummingbirds pump softened food from their crops down the throats of their offspring.

The activity of the parents at this season is amazing. Think of the day's work before a pair of Chickadees with a family of six or eight fledglings clamoring for food from daylight to dark!

But the young birds themselves furnish far more interesting and valuable subjects for study. None of the higher animals can be reared so easily without the aid of a parent. We therefore can not only study their growth of body and mind when in the nest and attended by their parents, but we can isolate the young of præcocial birds, such as Chickens, from other birds and study their mental development where they have no opportunity to learn by imitation. In this way students of instinct and heredity have obtained most valuable results.*

* Read Lloyd Morgan's Habit and Instinct (Edward Arnold, New York city).

CHAPTER VII.

HOW TO IDENTIFY BIRDS.

The preceding outline of the events which may enter into a bird's life-history has, I trust, given some idea of the possibilities attending the study of birds in the field. We come now to the practical question of identification. How are we to find birds, and, having found them, how are we to learn their names?

From April to August there is probably not a minute of the day when in a favorable locality one can not see or hear birds; and there is not a day in the year when at least some birds can not be found. In the beginning, therefore, the question of finding them is simply a matter of looking and listening. Later will come the delightful hunts for certain rarer species whose acquaintance we may make only through a knowledge of their haunts and habits.

Having found your bird, there is one thing absolutely necessary to its identification : *you must see it definitely.* Do not describe a bird to an ornithologist as "brown, with white spots on its wings," and then expect him to tell you what it is. Would you think of trying to identify flowers of which you caught only a glimpse from a car window in passing? You did not see them definitely, and at best you can only carry their image in your mind until you have opportunity to see them in detail.

So it is with birds. Do not be discouraged if the books fail to show you the brown bird with white spots on its wings. Probably it exists only through your hasty observation.

Arm yourself with a field- or opera-glass, therefore, without which you will be badly handicapped, and look your bird over with enough care to get a general idea of its size, form—particularly the form of the bill—color, and markings. Then—and I can not emphasize this too strongly—put what you have seen into your note-book *at once*. For, as I have elsewhere said, "not only do our memories sometimes deceive us, but we really see nothing with exactness until we attempt to describe it."

It is true that all the birds will not pose before your glasses long enough for you to examine them at your leisure, but many of them will, and in following the others you will have all the excitement of the chase. Who knows what rare species the stranger may prove to be!

From your description, and what added notes on voice and actions you may obtain, the field key and illustrations on the succeeding pages should make identification a simple matter.* You should also take into consideration the season of the year when a bird is present, and not call a summer bird by a winter bird's name. The dates of migration given in the following pages will be of assistance here. They refer to the vicinity of New York city, where, in the spring, birds arrive about a week later

---

* The publishers' liberality has resulted in securing bird portraits of unusual excellence. Mr. Seton Thompson is an ornithologist as well as an artist; his subjects are personal friends. He has spared no effort to make these pictures characteristic life sketches, and I venture to claim that, as a whole, they excel in truth and beauty any bird-drawings ever published in this country.

The descriptions accompanying these plates are designed to give only what the plates lack—that is, color, the *pattern* of coloration being clearly indicated by the drawing itself.

than in central Illinois or at Washington, D. C., and a week earlier than at Boston. In the fall these conditions are reversed.

A Bird's Biography.—As a further guide to your observation a list of the principal details which enter into a bird's life-history is appended:

1. DESCRIPTION (of size, form, color, and markings).
2. HAUNTS (upland, lowland, lakes, rivers, woods, fields, etc.).
3. MOVEMENTS (slow or active, hops, walks, creeps, swims, tail wagged, etc.).
4. APPEARANCE (alert, pensive, crest erect, tail drooped, etc.).
5. DISPOSITION (social, solitary, wary, unsuspicious, etc.).
6. FLIGHT (slow, rapid, direct, undulating, soaring, sailing, flapping, etc.).
7. SONG (pleasing, unattractive, continuous, short, loud, low, sung from the ground, from a perch, in the air, etc.; season of song).
8. CALL-NOTES (of surprise, alarm, protest, warning, signaling, etc.).
9. SEASON (spring, fall, summer, winter, with times of arrival and departure, and variations in numbers).
10. FOOD (berries, insects, seeds, etc.; how secured).
11. MATING (habits during courtship).
12. NESTING (choice of site, material, construction, eggs, incubation).
13. THE YOUNG (food and care of, time in the nest, notes, actions flight).

From observations of this kind, consisting of a simple statement of facts, you may philosophize according to your nature on the relation between habit and structure, colors and haunts, and intelligent adaptation to new conditions. Beware, however, lest you be led to draw faulty conclusions from insufficient observation. Do not make the individual stand for its species, or the species for its family, and remember that one is warranted in theorizing only when the facts in the case are facts indeed.

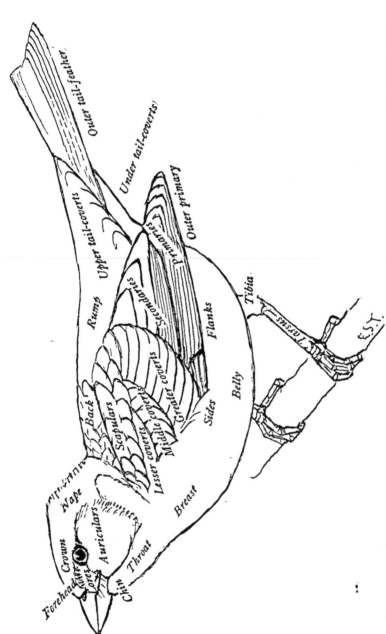

FIG. 25.—Topography of a bird. (House Sparrow, nearly natural size.) The "upper parts" include all the upper surface except wings and tail; the "under parts" all the under surface except wings and tail. The "length" is the distance from tip of the bill to end of the tail.

CHAPTER VIII.

A FIELD KEY TO OUR COMMON LAND BIRDS.

WHEN you have seen a bird with sufficient definiteness to describe its color, form, and actions, reference to the following key will often prove a short cut to its identity. This key is based only on adult males, who, because of their song, often brighter colors, and greater activity, are far more frequently observed than the females. But, knowing the male, you will rarely, during the nesting season, be at loss to recognize his mate.

In order to simplify the key, the water birds, Hawks, and Owls are omitted, in the belief that they will be more readily identified by reference to the plates.

The use of the key may be illustrated by the following example: Let us imagine that you see a Chipping Sparrow (Plate XLV) feeding about your doorstep. You note his size, chestnut cap bordered by white, black bill, brownish, streaked back, and grayish white, unmarked under parts. Turning now to the key, you will see that by exclusion the bird belongs in " Section V " of the " Third Group," and that it should be placed in subsection " 1 " of this section, which includes birds having the " under parts white or whitish, all *one* color, *without* streaks or spots." You have now two subdivisions to choose from—" A. Back *without* streaks or spots," and " B. Back brownish, streaked." Your bird falls under " B," where again you have two subdivisions, " *a*. Crown rufous. or chestnut, without streaks," and " *b*. Crown not rufous or chestnut." Your bird should be referred to " *a*," where you will at once find it described under " *a'* " as the Chipping Sparrow.

It should be borne in mind that living birds do not look as long as they really are. The measurement of "length" is taken with the bird's neck and tail outstretched in a straight line. In life the tail may be raised or drooped, while the neck is drawn in or curved, making the bird look shorter than measurement shows it to·be. Remember that the Robin measures ten inches, the House Sparrow six and one-fourth inches, and the Ruby-throated Humming-bird three and three-fourths inches in length, and you will have a basis for comparison.

FIRST GROUP.

BIRDS THAT FEED ON THE WING FOR LONG INTERVALS OF TIME WITHOUT PERCHING.

(Swallows, Swift, Nighthawk, Whip-poor-will.)

I. Size large, length over 9·00 inches: the spread wings over 15·00 inches in extent; generally seen only late in the afternoon or at dusk.

  1. A bird of the air, flying high, often over housetops in cities: a conspicuous white spot in each wing; note, a loud, nasal *peent*; sometimes dives earthward with a *booming* sound; May to Oct. . NIGHTHAWK, page 144.

  2. Haunts, near the ground, not often observed while feeding; call, given from a rock, stump, or similar perch, *whip-poor-will*, vigorously repeated; Apl. 25 to Oct. . . . . WHIP-POOR-WILL, page 146.

II. Size smaller, length under 9·00 inches; the spread wings less than 15·00 inches in extent; may be seen at any time of the day.

  1. Plumage entirely black.

    a. Length 5·50 inches; plumage sooty black; usually nests in chimneys; Apl. 25 to Oct.. . . . . CHIMNEY SWIFT, page 146.

    b. Length 8·00 inches; glossy, bluish black; nests in gourds or houses erected for its use; Apl. 25 to Sept. . . . PURPLE MARTIN, page 216.

  2. Plumage not entirely black; Apl. to Oct. . SWALLOWS, pages 214, 215.

SECOND GROUP.

CLIMBING AND CREEPING BIRDS.

(Nuthatches, Creepers, Woodpeckers.)

I. Birds *without* stiffly pointed tail-feathers, that climb either up or down.

  1. Length 6·00 inches; back gray, cap black, cheeks and under parts white; note, a nasal *yank, yank*; a permanent resident.

WHITE-BREASTED NUTHATCH, page 250.

2. Length 4.50 inches; back gray, cap black, a blackish streak through the face ; under parts reddish brown ; note, high and thin, like the tone of a penny trumpet: Sept. to Apl. RED-BREASTED NUTHATCH, page 25.

3. Length 5.25 inches; upper parts streaked black and white ; note, a thin wiry *see-see-see-see* ; Apl. 25 to Oct.

BLACK AND WHITE WARBLER, page 226.

II. Birds with stiffly pointed tail-feathers, that always climb upward.

1. Length 5.65 inches; plumage dull brown and black ; size small, bill slender ; an inconspicuous bird who winds his way up the trunks searching for insects' eggs, etc. ; note, fine and squeaky ; Sept. 25 to Apl.

BROWN CREEPER, page 246.

2. Plumage with more or less white, size larger, bill stouter, chisel-like, often used in hammering.

A. Length 9.75 inches; head red, back black ; flight showing a large white patch in the wing . . . RED-HEADED WOODPECKER, page 140.

B. Length 12.00 inches; crown gray ; a red band on the nape; flight showing a white patch on the lower back and yellow in the wings ; often flushed from the ground; note, *kee-yer* . . FLICKER, page 138.

C. Length 6.75 inches ; crown black ; back and wings black and white ; note, a sharp *peek* . . . . DOWNY WOODPECKER, page 138.

THIRD GROUP.

BIRDS NOT INCLUDED IN THE PRECEDING GROUPS.

(*Blackbirds, Orioles, Sparrows, Vireos, Warblers, Thrushes, etc.*)

SECTION I. With yellow or orange in the plumage.

SECTION II. With red in the plumage.

SECTION III. With blue in the plumage.

SECTION IV. Plumage conspicuously black, or black and white.

SECTION V. Birds not included in the preceding sections.

I. With yellow or orange in the plumage.

1. Throat yellow.

A. Throat and breast pure yellow, *without* streaks or spots.

 a. Length 5.10 inches ; cap, wings, and tail black ; back yellow ; song canarylike, sometimes uttered on the wing ; flight undulating, frequently accompanied by the notes *chic-o-ree, per-chic-o-ree* ; a permanent resident . . . . AM. GOLDFINCH, page 198.

 b. Length 5.95 inches ; lower belly and wing-bars white ; back olive-green ; frequents the upper branches, generally in woodland ; actions deliberate ; song loud and musical, uttered slowly, often with pauses : " See me ? I'm here ; where are you ? "; May to Sept.

YELLOW-THROATED VIREO, page 222.

 c. Length 5.25 inches ; cheeks and forehead black, bordered by ashy ; upper parts olive-green ; no wing-bars ; haunts thickets and undergrowth : movements nervous and active ; call-note *pit* or *chack ;* song, a vigorous, rapid *witch-e-wee-o, witch-e-wee-o* ; May to Oct.

MARYLAND YELLOW-THROAT, page 234.

d. Length 7·45 inches; upper parts olive-green; no wing-bars; a white
line before the eye; haunts thickets and undergrowth; song, a striking mixture of whistles, *chucks*, and *caws*, sometimes uttered on the
wing; May to Sept. . . . . . . YELLOW-BREASTED CHAT, page 236.

B. Under parts streaked with reddish brown; length 5·00 inches; general appearance of a yellow bird; haunts shrubbery of lawns, orchards,
second growths, and particularly willows near water; song, rather loud,
wĕĕ, chĕĕ-chĕĕ-chĕĕ, chĕr-wĕĕ, or *chĕĕ-chĕĕ-chĕĕ-chĕĕ, wày-o*; May to Sept.
YELLOW WARBLER, page 228.

C. Breast yellow, with a conspicuous black crescent; length 10·75 inches;
haunts fields and meadows, largely terrestrial; flight quail-like, outer
tail-feathers white, showing when on the wing; song, a loud, musical
whistle; a permanent resident . . . . . . MEADOWLARK, page 174.

2. Throat white.

A. With yellow on the sides.

a. Length 5·50 inches; rump yellow; breast streaked or spotted with
black; tail-feathers marked with white; note, a characteristic *tchip*;
Sept. to May, usually rare or local in winter.
MYRTLE WARBLER, page 228.

b. Length 5·00 inches; no black on under parts or white in the tail; yellow extending along the whole sides; back olive-green, iris white;
haunts thickets; call, an emphatic " Who are you, eh ? "; May to Oct.
WHITE-EYED VIREO, page 222.

c. Length 5·25 inches; tail and wings banded with yellow, showing
conspicuously in flight; haunts woodland; movements active, much
in the air, tail frequently spread; May to Oct. REDSTART, page 230.

B. No yellow on sides.

a. Length 6·75 inches; a yellow line from the bill to the eye; crown
black, with a white stripe through its center; haunts in and about
thickets and bushy woodlands; song, a high, clear, musical whistle;
call-note, *chink* . . . . . WHITE-THROATED SPARROW, page 188.

b. Length 4·00 inches; a yellow, or yellow and orange crown-patch, bordered by black; flits restlessly about outer limbs of trees and bushes;
note, a fine *ti-ti*; Oct. to Apl. GOLDEN-CROWNED KINGLET, page. 251.

3. Throat neither yellow nor white.

A. Length 12·00 inches; white rump and yellow in wings showing conspicuously in flight; a black breast-band; note, a loud *kèe-yer*.
FLICKER, page 140.

B. Length 9·00 inches; crested; breast ashy, belly yellow; tail-feathers
largely pale brownish red; haunts upper branches in woodland; note,
a loud questioning or grating whistle; May to Sept.
CRESTED FLYCATCHER, page 152.

C. Length 7·50 inches; throat and head black; breast, belly, and lower
back deep orange; haunts fruit and shade trees; song, a loud, ringing
whistle; May to Sept. . . . . . . BALTIMORE ORIOLE. page 164.

D. Length 7·20 inches; crested; grayish brown; a black line through
the eye; tail tipped with yellow; generally seen in small flocks; note
thin and weak . . . . CEDAR WAXWING, page 216.

II. With red in the plumage.

1. With red on the under parts.

A. Throat red.

a. Length 7·25 inches; wings and tail black; rest of plumage bright scarlet; call-note, *chip-chirr*; May to Sept.

SCARLET TANAGER, page 211.

b. Length 6 20 inches; dull. pinkish red, wings and tail brownish; frequently seen feeding on buds or blossoms; call-note, a sharp *chink*, often uttered during flight; song, a sweet, flowing warble.

PURPLE FINCH, page 200.

c. Length 6·20 inches; dull red or green tinged with red; mandibles crossed; generally seen in flocks; feeds on pine cones.

AM. CROSSBILL, page 196.

d. Length 5·30 inches; a red crown-cap: back streaked black and brown; breast rosy; feeds on seeds or catkins; Nov. to Mch.

REDPOLL, page 194.

B. Throat black.

a. Length 8·00 inches; breast rose-red, rest of plumage black and white; song loud and musical; call-note, *peek*; May to Sept.

ROSE-BREASTED GROSBEAK, page 202.

b. Length 8 00 inches; a conspicuous crest; region about the base of the bill black; rest of the plumage and bill red; song, a clear whistle; resident from New York city southward.

CARDINAL, page 207.

c. Length 5·40 inches; wings and tail banded with orange-red, showing conspicuously in flight: movements active; much in the air; tail frequently spread; haunts woodland; May to Oct.

REDSTART, page 230.

2. No red on the under parts.

A. Length 9·50 inches; black; shoulders red; haunts marshes; migrates in flocks; Mch. to Oct. . . . . RED-WINGED BLACKBIRD, page 166.

B. Length 5 25 inches: crown-cap red; chin black; rest of under parts streaked with blackish; feeds on seeds and catkins; Nov. to Mch.

REDPOLL (im), page 194.

C. Length 4·00 inches; under parts whitish; back olive-green; a ruby crown-patch; eye-ring white; movements restless, wings flitted nervously; call-note, *cack*; song remarkably loud and musical; Sept. and Oct.; Apl. and May . . . . . RUBY-CROWNED KINGLET, page 252.

III. With blue in the plumage.

A. Length 11·50 inches; a conspicuous crest; upper parts dull blue; under parts whitish; a black patch on the breast.

BLUE JAY, page 163.

B. Length 7·00 inches; upper parts bright blue; under parts cinnamon-brown . . . . BLUEBIRD, page 260.

C. Length 5 50 inches; entire plumage indigo-blue; May to Oct.

INDIGO BUNTING, page 206.

D. Length 13·00; bluish gray; haunts near water; feeds on fish, which it catches by darting on them at the surface . KINGFISHER, page 136.

IV. Plumage conspicuously black, or black and white.

1. Black and white birds.

   A. Throat black.

      a. Length over 6 00 inches.

         a^1. Entire under parts black; nape buffy; rump white; a musical dweller of fields and meadows; frequently sings on the wing; May to Sept. . . . . BOBOLINK, page 170.

         a^2. Breast rose-red; rest of the plumage black and white; song rapid, loud and musical; call-note, *peek*; a tree dweller in rather open woodland; May to Sept.

                              ROSE-BREASTED GROSBEAK, page 202.

         a^3. Sides reddish brown; rest of the plumage black and white; call-note, *chewink* or *towhèe*; inhabits the undergrowth; often seen on ground scratching among fallen leaves; Apl. 25 to Oct.

                                    TOWHEE, page 204.

      b. Length under 6·00 inches.

         b^1. Crown black; cheeks white; back ashy; unstreaked; call, *chick-a-dee*, or a musical, double-noted whistle; a permanent resident.

                                  CHICKADEE, page 248.

   B. Throat and other parts white or whitish.

      a. Length 8·50 inches; upper parts blackish slate-color; tail tipped with white; a bird of the air, catching its insect food on the wing, and occasionally sallying forth from its exposed perch in pursuit of a passing Crow; note, an unmusical, steely chatter; May to Sept.

                                  KINGBIRD, page 150.

      b. Length 6·90 inches: upper parts washed with rusty; generally seen in flocks: terrestrial; Nov. to Mch. . . . . SNOWFLAKE, page 196.

2. No white in the plumage.

   A. Length 19·00 inches; jet black . . . . . . AM. CROW, page 161.

   B. Length 12·00 inches; black with metallic reflections; iris yellowish; migrates in flocks; nests usually in colonies in coniferous trees; voice cracked and reedy; tail "keeled" in short flights; a walker; Mch. to Nov. . . . . PURPLE GRACKLE, page 168.

   C. Length 9·50 inches; shoulders red; haunts marshes; call, *kong-quĕr-rēè*; Mch. to Oct. . . . . . . RED-WINGED BLACKBIRD, page 166.

   D. Length 7·90 inches; head and neck coffee-brown; frequently seen on the ground near cattle; Mch. to Nov. . . . . . COWBIRD, page 176.

V. Birds not included in the preceding sections (that is, plumage without either yellow, orange, red, or blue; not conspicuously black, or black and white).

1. Under parts white or whitish, all *one* color, *without* streaks or spots.

   A. Back *without* streaks or spots.

      a. Back olive-green; gleaners, exploring the foliage for food or flitting about the outer branches.

         a^1. Length 6·25 inches; a white line over the eye bordered by a narrow black one; cap gray: iris red; song, a rambling recitative: "You see it—you know it—do you hear me?" etc.; May to Oct.

                             RED-EYED VIREO, page 221.

a². Length 5·75 inches; a white line over the eye not bordered by black; prefers the upper branches of rows of elms and other shade trees; song, a rich, unbroken warble with an alto undertone; May to Sept. . . . . WARBLING VIREO, page 222.

a³. Length 4·00 inches; no white line over the eye; eye-ring and wing-bars white; a tiny, unsuspicious bird; flits about the outer branches of trees and shrubs; wings twitched nervously; note, *cack*; song, a remarkably loud, musical whistle; Sept. and Oct.; Apl. and May . . . . . . RUBY-CROWNED KINGLET, page 252.

b. Back olive-green or dusky olive; flycatchers who capture their prey on the wing by darting for it, and while perching are quiet and erect.

b¹. Length 7·00 inches; frequently found nesting under bridges or about buildings; crown blackish; tail wagged nervously; notes, *pee, pee*, and *pewit-phœbe*; Mch. to Oct. . . . PHŒBE, page 154.

b². Length 6·50 inches; haunts wooded growths; note, a plaintive *pee-a-wee*; May to Sept. . . . . . . WOOD PEWEE, page 158.

b³. Length 5·40 inches; haunts orchards, lawns, and open woodlands; note, *chebéc, chebéc* . . . . . . LEAST FLYCATCHER, page 156.

c. Back gray or bluish gray.

c¹. Length 6·50 inches; a gray, crested bird; forehead black; no white in the tail; note, a whistled *peto, peto*, or hoarse *de-de-de-de*; resident from New York city southward . . TUFTED TIT, page 250.

c². Length 8·50 inches; a white band at the end of the tail; a concealed orange-red crest; a bird of the air, catching its insect food on the wing, and occasionally sallying forth from its exposed perch in pursuit of a passing Crow; note, an unmusical, steely chatter; May to Sept. . . . . KINGBIRD, page 150.

d. Back brown.

d¹. Length 5·00 inches; a nervous, restless, excitable bird; tail often carried erect; song sweet, rapid and rippling, delivered with *abandon*; Apl. 25 to Oct. . . . . HOUSE WREN, page 240.

d². Length 12·25 inches; slim, brownish birds with long tails; flight short and noiseless; perch *in* a tree, not in an exposed position; note, *tut-tut, cluck-cluck*, and *cow-cow*; May to Oct.

YELLOW-BILLED CUCKOO, BLACK-BILLED CUCKOO, page 132.

B. Back brownish, streaked.

a. Crown rufous or chestnut without streaks.

a¹. Length 5·25 inches; bill black; a whitish line over the eye; a familiar bird of lawns and door-yards; song, a monotonous *chippy-chippy-chippy*; Apl. to Nov. . . CHIPPING SPARROW, page 186.

a². Length 5·70 inches; bill *reddish brown*, back rufous or rufous-brown; wing-bars and eye-ring whitish; haunts dry, bushy fields and pastures; song, a musical, plaintive *cher-wee, cher-wee, cher-wee, cheeo, dee-dee-dee-dee*; Apl. to Nov. FIELD SPARROW, page 182.

a³. Length 5·90 inches; forehead black; crown and wings chestnut-rufous; flanks pale grayish brown; haunts marshes; song, a rapidly repeated *weet-weet-weet*, etc.; Mch. to Nov.

SWAMP SPARROW, page 180.

7

    b. Crown not rufous or chestnut.

        b¹. Length 6·75 inches; crown blackish, with a central whitish stripe; throat white; breast gray; a yellow spot before the eye; haunts in and about thickets and bushy woodlands; song, a high, clear, musical whistle; call-note, *chink*.

                        WHITE-THROATED SPARROW, page 188.

        b². Length 5·20 inches; bill slender; a white line over the eye; tail carried erect; haunts reedy marshes; call-note scolding; song rippling; May to Oct.

                        LONG-BILLED MARSH WREN, page 244.

2. Under parts white or whitish, *streaked* or *spotted.*

  A. Back streaked.

    a. Length 6·10 inches; outer tail-feathers white, showing conspicuously when the bird flies; haunts dry fields and roadsides; song musical; Apl. to Nov. . . . . VESPER SPARROW, page 184.

    b. Outer tail-feathers *not* white.

        b¹. Length 6·30 inches; breast with numerous spots tending to form one large spot in its center; haunts on or near the ground, generally in the vicinity of bushes; call-note, *chimp*; song musical; a permanent resident . . . . SONG SPARROW, page 178.

        b². Length 6·35 inches; breast grayish with *one* spot in its center; Oct. to Apl. . . . . TREE SPARROW, page 194.

  B. Back *not* streaked; breast spotted.

    a. Length 11·40 inches; tail 5·00 inches; wing-bars white; upper parts, wings, and tail bright cinnamon-brown; haunts undergrowth; sings from an exposed and generally elevated position; song loud, striking, and continuous; Apl. 25 to Oct. . . BROWN THRASHER, page 240.

    b. Length under 9·00 inches; tail under 3·00 inches; no wing-bars; back *reddish* or *cinnamon-brown.*

        b¹. Length 8·25 inches; breast and *sides* heavily marked with large, *round*, black spots; head and upper back *brighter* than lower back and tail; call-note, a sharp *pit* or liquid *quirt*; May to Oct.

                        WOOD THRUSH, page 256.

        b². Length 7·15 inches; breast with wedge-shaped black spots; sides *unspotted*, washed with *brownish ashy*; tail reddish brown, *brighter* than back; call-note, a low *chuck*; Apl. 10 to May 10; Oct. and Nov. . . . . HERMIT THRUSH, page 258.

        b³. Length 7·50 inches; upper breast *lightly* spotted with small, wedge-shaped, brownish spots; tail the same color as the back; sides *white*; call-note, a clearly whistled *wheèu*; May to Sept.

                        WILSON'S THRUSH, page 254.

    c. Length under 9·00 inches; tail under 3·00 inches; no wing-bars; back *olive-green.*

        c¹. Length 6·10 inches; center of crown pale brownish bordered by black; haunts on or near the ground in woodland; a *walker*; song, a ringing crescendo, teacher, *teacher*, TEACHER, TEACHER, *TEACHER*; May to Sept.. . . . . . . OVEN-BIRD, page 232.

3. Under parts *not* white or whitish, all *one* color, *without* streaks.

   A. Length 8·50 inches; slate-color: cap and tail black; inhabits the lower growth; call-note, nasal; song highly musical and varied; Apl. 25 to Oct. . . . . Catbird, page 237.

   B. Length 7·20 inches; grayish brown; conspicuously crested; a black line through the eye; tail tipped with yellow; generally seen in small flocks; note thin and weak . . . . . Cedar Waxwing, page 216.

   C. Length 5·50 inches; under parts cream-buff; a conspicuous whitish line over the eye; upper parts reddish brown; movements active; tail carried erect; haunts lower growth; notes loud and striking; resident from New York city southward . . . . Carolina Wren, page 244.

4. Throat and upper breast black or slate-color, very different from the white or chestnut belly.

   A. Throat black.

     a. Belly and rump chestnut; head, wings, and tail black; length 7·30 inches; haunts orchards and shade trees; song highly musical; May to Sept. . . . . Orchard Oriole, page 166

     b. Belly white; sides reddish brown; tail black and white; length 8·35 inches; haunts undergrowths; call-note, *chewink* or *towhee*; Apl. 25 to Oct. . . . . Towhee, page 204.

   B. Throat slate-color.

     a. Back and wings slate-color; outer tail-feathers and belly white; length 6·25 inches; haunts generally on or near the ground about shrubbery; Oct. to Apl. . . . . Junco, page 192.

5. Throat streaked with black and white; rest of under parts reddish brown; upper parts grayish slate-color; length 10·00 inches . Robin, page 260.

THE WATER BIRDS.

DIVING BIRDS. (ORDER PYGOPODES.)

GREBES. (FAMILY PODICIPIDÆ.)

THE study of water birds requires special advantages and equipments, among which are a suitable location, much time, and a gun. Our coasts and shores are becoming so popular as "resorts" that many of the former haunts of waterfowl are now thickly populated, and the birds are comparatively rare. Furthermore, the larger number of our water birds nest in the far North and winter in the South, visiting the Middle States only while on their migrations. It is evident, therefore, that if we would become familiar with these birds, we must devote ourselves especially to their pursuit.

There are, however, some species, notably those which frequent bodies of fresh water and nest in this latitude,

Pied-billed Grebe,
Podilymbus podiceps.
Plate II.

which deserve to be ranked among our commoner birds. Of these, one of the best known, by name at least, is the Pied-billed Grebe, whose aquatic powers have given it the expressive title of Hell-diver.

Under favorable conditions this little Grebe may breed anywhere from the Argentine Republic to British America, but in the Middle States it occurs chiefly as a spring and fall migrant. When nesting, a quiet, reedy pond or lake is chosen for a home, the nest being made on a pile of decaying vegetation. The eggs, four to eight in number, are dull white, more or less stained by the nesting material, which the parent bird rarely fails to place over

PLATE II.

PIED-BILLED GREBE.

Length, 13·50 inches. *Summer plumage*, upper parts blackish brown ; throat and spot on bill black ; fore neck brownish, rest of under parts grayish white. *Winter plumage*, similar, but without black on throat or bill.

them when leaving the nest. The young are born covered with down and can swim at birth. The Pied-billed Grebe is one of our most aquatic birds. When pursued, it prefers diving to flight, and the marvelous rapidity with which it can disappear from the surface of the water, to reappear in a quite unexpected place, justifies its reliance on its own natatorial powers. It can swim under water with only its bill exposed, when it becomes practically invisible.

When on land Grebes progress awkwardly. They can, it is said, stand erect on their toes, but, when resting, support themselves on the whole length of the foot or tarsus (see Fig. 8, the Great Auk).

On the wing Grebes resemble small Ducks, but their pointed bill and their feet stretched out behind the rudimentary tail will serve to distinguish them.

Loons. (Family Urinatoridæ.)

The Loon, like its small relative the Grebe, is known to almost every one by name, but only those who have visited its summer haunts among the Northern lakes and heard its wild call can be said to know it. Nuttall writes

Loon,
Urinator imber.
Plate III.

of its cry as "the sad and wolfish call of the solitary Loon, which, like a dismal echo, seems slowly to invade the ear, and, rising as it proceeds, dies away in the air." It "may be heard sometimes for two or three miles, when the bird itself is invisible, or reduced almost to a speck in the distance." The Loon is as aquatic in habits as the Grebe, but is much stronger on the wing. It migrates by day, and probably also by night, and we may sometimes see it passing over—a large, ducklike bird—in March and October.

When on land, it is nearly helpless, progressing awk-

PLATE III.

LOON.

Length, 32 00 inches. *Summer plumage*, upper parts and fore neck black and white ; breast and belly white. *Winter plumage*, upper parts dark grayish ; under parts white.

87

wardly by the use of feet, wings, and bill. For this reason it nests near the water's edge, often where it can slide from the eggs directly into its true element. The nest is a slight depression in the earth, in which are laid two elliptical eggs, in color olive-brown, slightly spotted with blackish.

LONG-WINGED SWIMMERS. (ORDER LONGIPENNES.)

GULLS AND TERNS. (FAMILY LARIDÆ.)

No birds are more widely distributed than the Gulls and Terns. Some species are pelagic, visiting the land only at long intervals and when nesting; others live along the coast, and several species resort to inland waters. About one hundred species are known,

Herring Gull,
Larus argentatus smithsonianus.
Plate IV.

fifty being Gulls and fifty Terns. The former are, as a rule, larger, stouter birds than the latter, and, generally speaking, are more maritime. The commonest of the ten species found in the Eastern States is the Herring Gull. It nests from Maine northward, and is found southward along our coast from October 1 to April. This is the Gull we see in such numbers in our bays and harbors, flying gracefully and apparently aimlessly about, but in reality ever keeping its bright black eyes fixed on the water in search of some floating morsel, which it deftly picks from the surface. It frequently follows vessels, hanging over the stern day after day, and deserting its post only to feed on scraps thrown overboard from the galley. There are said to be reliable records of these birds following the same vessel from the Irish coast to New York Harbor.

Gulls do excellent service in devouring much refuse that would otherwise be cast ashore to decay; but, useful

PLATE IV.

HERRING GULL.

Length, 24·00 inches. *Adult*, back and wings pearl-gray ; end of primaries marked with black ; rest of plumage white. *Young*, dark grayish, primaries and tail brownish black.

PETRELS.

Length, 7·50 inches. Black, upper tail-coverts white.

as they are as scavengers, I feel that their place in Nature
is to animate the barren wastes of the sea. How, when
at sea, the presence of a single Gull changes the whole
aspect of Nature! The great expanse of water, which
before was oppressive in its dreary lifelessness, is transformed by the white-winged Gulls into a scene of rare
beauty. Every voyager, be he naturalist or not, admires
their grace of form and motion. They seem born of the
waves, and as much a part of the ocean as the foamy
whitecaps themselves.

The beautiful Terns or Sea Swallows are even more
graceful than the Gulls. They are slenderer birds, lighter

Common Tern, and more active on the wing, with long,
Sterna hirundo. forked tails and pointed bills. They
Plate X. arrive from the South in May and re-
main until September, nesting in colonies.

Terns are littoral rather than pelagic, seldom being
found far from the shore. Like the Gulls, they seem so
in harmony with their surroundings that no coast view is
perfect from which the Terns are missing. They add
the requisite touch of life, and make still more impressive
the thunder of the surf dashing over rocks or curling in
long, combing waves on the beach.

During recent years these birds have been killed in
such numbers for millinery purposes that on the middle
Atlantic coast the only survivors exist on three or four
uninhabited islets. If one protests against the merciless
destruction of these exquisite creatures the excuse is,
"Well, what good are they?"—an answer betraying such
an utter lack of appreciation of beauty that explanation
seems hopeless. But can we not learn, before it is too
late, that these birds are even more deserving of protection than the works of art we guard so zealously?

TUBE-NOSED SWIMMERS. (ORDER TUBINARES.)

PETRELS. (FAMILY PROCELLARIIDÆ.)

Petrels, or "Mother Carey's Chickens," are true children of the sea. Their home is the ocean, and they come to land only when nesting. To the landsman, therefore, they are strangers,

Petrels,
Plate IV.

but to most people who have been to sea they are known as the little, white-rumped swallow-like birds who on tireless wing follow in the wake of the ship day after day, patiently waiting for the food which experience tells them will be thrown overboard.

Two species of Petrels are found off our coasts, Wilson's and Leach's. The former has a yellow area in the webs of the toes and a square tail, while Leach's Petrel has the webs of the toes wholly black and a slightly forked tail. These differences, however, would not be appreciable at a distance. Wilson's Petrel nests in certain islands of the southern hemisphere in February, and later migrates northward, reaching our latitude in May and spending the summer, or what in fact is its winter, in the North Atlantic. It is, therefore, probably the Petrel most frequently seen by transatlantic voyagers at this season.

Leach's Petrel nests on our coasts from Maine northward, arriving from the South in May. The nest is made in a burrow in the ground or beneath a rock, and a single white egg is laid. Generally one of the birds spends the day on the nest while its mate is at sea, but at night the incubating bird leaves the nest, its place being taken probably by the one who has been feeding during the day.

LAMELLIROSTRAL SWIMMERS. (ORDER ANSERES.)

Ducks, Geese, and Swans. (Family Anatidæ.)

This family contains some two hundred species, and is represented in all parts of the world. It includes five subfamilies : the Mergansers (*Merginæ*), or Fish-eating Ducks; the Pond or River Ducks (*Anatinæ*), the Bay or Sea Ducks (*Fuligulinæ*); the Geese (*Anserinæ*); and the Swans (*Cygninæ*).

Ducks, like all hunted birds, are exceedingly wild, and comparatively few species will come within reach of the student's opera-glass. The group may therefore be reviewed briefly. The Mergansers or Shelldrakes, numbering three species, have narrow, serrate bills which enable them to hold the fish they pursue and catch under water (see Fig. 18).

The River Ducks have little or no lobe or flap on the hind toe. In this group belong our Mallard, Widgeon, Pintail, Blue-winged and Green-winged **River Ducks,** Teals, Black Duck, Wood Duck, and **Plate V.** others. All but the last two nest in the North and are found in our latitude only during their spring and fall migrations, or, if the weather be mild, in the winter. The Black Duck and Wood Duck nest rarely in the Middle States.

All these birds feed in shallow water by " dabbling " or " tipping," terms which will be readily understood by any one who has watched domesticated Ducks feeding.

The Bay and Sea Ducks, on the contrary, are divers, and may descend to the bottom in water more than one hundred and fifty feet in depth. They are to be distinguished from the River Ducks by the presence of a flap or lobe on the hind toe. The commoner members of

PLATE V.

1 WOOD DUCK.　　4 GREEN-WINGED TEAL.
2 PINTAIL.　　　 5 BLUE-WINGED TEAL.
3 MALLARD.　　　 6 CANADA GEESE.

93

this subfamily are the Redhead, Canvasback, Scaup or Broadbill, Whistler, Bufflehead, Old Squaw, Eider, three species of Scoters or " Coots " and Ruddy Duck. These are all northern-breeding birds who visit the waters of our bays and coasts during their migrations or in the winter.

The bill in both River and Bay Ducks has a series of gutters on either side which serve as strainers. The birds secure a large part of their food—of small mollusks, crustaceans, and seeds of aquatic plants—from the bottom, taking in with it a quantity of mud, which they get rid of by closing the bill and forcing it out through the strainers, the food being retained.

Geese are more terrestrial than Ducks, and, though they feed under water by tipping, often visit the land to procure grass, corn, or cereals, which they readily nip off. The white-faced, black-necked Canada Goose is our only common species. Its long overland journeys, while migrating, render it familiar to many who have seen it only in the air. It migrates northward in March and April and returns in October and November, breeding from the Northern States northward and wintering from New Jersey southward.

The two Swans, Whistling and Trumpeter, found in North America, are generally rare on the Atlantic coast.

HERONS, STORKS, IBISES, ETC. (ORDER HERODIONES.)

HERONS AND BITTERNS. (FAMILY ARDEIDÆ.)

OF the seventy-five known members of this family fourteen inhabit eastern North America. Most of these are Southern in distribution, only six or seven species regularly visiting the Northern States. Their large size renders Herons conspicuous, and, though worthless as food, few so-called sportsmen can resist the temptation of shooting at them when opportunity offers. Several of the Southern species, notably the Snowy Heron and White Egret, are adorned during the nesting season with the beautiful " aigrette " plumes which are apparently so necessary a part of woman's headgear that they will go out of fashion only when the birds go out of existence. One can not blame the plume hunters, who are generally poor men, for killing birds whose plumes are worth more than their weight in gold—the blame lies in another quarter. But I have no words with which to express my condemnation of the man who kills one of these birds wantonly.

The presence of a stately Great Blue Heron or " Crane " adds an element to the landscape which no

Great Blue Heron,
Ardea herodias.
Plate VI.

work of man can equal. Its grace of form and motion, emphasized by its large size, is a constant delight to the eye ; it is a symbol of the wild in Nature ; one never tires of watching it. What punishment, then, is severe enough for the man who robs his fellows of so pure a source of enjoyment ? A rifle ball turns this noble creature into a useless mass of flesh and feathers ; the loss is irreparable. Still, we have no law to prevent it. Herons are said to devour large numbers of small fish. But is not the laborer worthy of his hire ? Are the fish more valuable than this, one of the grandest of birds ?

The Great Blue Heron breeds throughout North America, but there are now only a few localities in the northeastern States where it may be found nesting. We usually see it, therefore, as a migrant in April and May, and from August to November.

The Little Green Heron is the smallest, as the Great Blue Heron is the largest, of our Herons. Its small size, preference for wooded regions instead of marshes, and habit of nesting alone, not in flocks, like most Herons, accounts for its being relatively common. It arrives from the South about April 20, and nests early in

Little Green Heron,
Ardea virescens.
Plate VI.

May. The nest, as is usual in this family, is a rude platform of sticks and is placed in a bush or the lower branch of a tree, often overhanging the water. The eggs number from three to six, and in color are pale greenish blue. The young, although born with a covering of hairlike feathers, are quite helpless and are reared in the nest. Adults have the crown and back dark, glossy green, the neck reddish brown.

The notes of this little Heron are a clear whistle and a harsh *squawk*, uttered when it is frightened. It then seeks refuge by alighting in a distant bush or tree, and with upstretched neck and twitching tail watches the intruder.

The Night Heron, or Squawk, doubtless owes its escape from the fate of most Herons to its nocturnal habits. These birds arrive from the South in April and remain until October. They nest in large colonies, a rookery not far from New York city being inhabited by at least one thousand

**Black-crowned
Night Heron,**
*Nycticorax nycticorax
nævius.*
Plate VI.

pairs. It is in a low, wooded tract, and the nests are built in the trees at an average height of thirty feet. The eggs number four to six, and in color are pale bluish green.

At night, while feeding, these Herons are doubtless distributed over a wide area. When flying, they often utter a loud *squawk*, the origin of one of their common names. It is a surprising sound when heard near by at night, and has doubtless aroused the curiosity of many persons who live near a line of flight followed by these birds in going to and from their nests.

In the Haunts of the Heron.

Ernest Seton Thompson

PLATE VI.

LITTLE GREEN HERON.
(*Length, 17·00 inches*)

GREAT BLUE HERON.
(*Length, 45·00 inches.*)

BLACK-CROWNED NIGHT HERON
(YOUNG AND ADULT).
(*Length, 24·00 inches.*)

97

8

The Bittern, or Stake Driver, is a summer resident of our larger marshes, arriving early in April and remaining **American Bittern,** until October. Though by no means *Botaurus lentiginosus.* common, its notes are so loud and re-Plate VII. markable that even a single calling bird is more likely to attract attention than many smaller abundant species. Under favorable circumstances these notes may be heard for at least three fourths of a mile. They are of two kinds. One is described as the " pumping " call, and is generally written *púmp-er-lunk*, *púmp-er-lunk*, *púmp-er-lunk*, while the other is deceptively like the sound produced by driving a stake in the mud. Mr. Bradford Torrey, one of the few ornithologists who has observed the bird while it was uttering these singular cries, tells us (The Auk, vi, 1889, p. 1) that they are attended by violent, convulsive movements of the head and neck, which suggest the contortions of a seasick person, but that the bird's bill is neither immersed in water nor plunged in the mud, as has been popularly supposed.

CRANES, RAILS, ETC. (ORDER PALUDICOLÆ.)

RAILS AND COOTS. (FAMILY RALLIDÆ.)

RAILS are marsh-inhabiting birds, more often heard than seen. They are very reluctant to take wing, and when pursued seek safety by running or hiding rather than by flying. When flushed, they go but a short distance, and with dangling legs soon drop back into the grasses.

Of the one hundred and eighty members of this family, fourteen inhabit North America and eight visit the northeastern United States. Only three or four of these, however, are abundant, the most numerous and

PLATE VII.

AMERICAN BITTERN

Length, 28·00 inches. A black streak on neck ; body brown and buff ; primaries slate-color.

SORA.

Length, 8·50 inches. *Adult*, upper parts olive-brown, black, and white ; throat and face black, breast slate, belly white, flanks black and white. *Young*, similar, but face, throat, and breast white, washed with brownish.

generally distributed species being our Sora or Carolina Rail, so well known to sportsmen. This bird passes

Sora,
Porzana carolina.
Plate VII.

us in the spring in April and nests from Massachusetts northward. It returns in August and lingers in our wild-rice marshes until October. During the nesting season it has two calls—a whistled, *ker-wee*, and a high, rolling *whinny*. In the fall it utters a *kuk* or *peep* when disturbed.

There is no sexual difference in color in this species, but birds of the year lack the black about the base of the bill and on the throat, and have the breast washed with cinnamon.

Our other species of Rail are the King, Yellow, and Little Black Rail, all of which are rare; the Virginia Rail,

Clapper Rail,
Rallus crepitans.
Plate VIII.

which is more common, and the Clapper Rail or Marsh Hen, an abundant species in some of the salt marshes along our coasts from Long Island southward. It is a noisy bird with a peculiar cackling call which it utters in a way that suggests the sound produced by some automatic toys.

Its nest is made of dried grasses, the surrounding marsh grass being slightly arched over it. Eight to twelve buffy, speckled eggs are laid, a number which, in connection with the abundance of the bird, has led to the persistent robbing of its nests by men who sell the eggs for food. As a result of this practice the birds have greatly decreased in numbers during recent years.

The Coot, Mud-hen, or Crow-duck differs from the Rails in having lobed toes (see Fig. 12) and in being

American Coot,
Fulica americana.
Plate VIII.

more aquatic. In fact, it is more like a Duck in habits than like a Rail, but its pointed, white-tipped bill will prevent its being mistaken for one.

PLATE VIII.

AMERICAN COOT.

Length, 15·00 inches. Head and neck blackish, body slate ; under tail-coverts, tips of secondaries, and end of bill white.

CLAPPER RAIL.

Length, 14·50 inches. Upper parts pale greenish olive and gray; throat white, breast pale cinnamon, flanks gray and white.

It rarely breeds on the Atlantic coast, but is sometimes common on our marsh-bordered streams in the
fall.

SHORE BIRDS. (ORDER LIMICOLÆ.)

SNIPES AND SANDPIPERS. (FAMILY SCOLOPACIDÆ.)

THE successful pursuit of shore birds on our coasts
requires a special knowledge of their notes and habits.
Thirty of the one hundred known species visit us annually, but of this number only two or three nest, most of
the others migrating in May to their breeding grounds in
the far North. The return migration takes place during
July, August, and September, but with some exceptions
these birds are seen only by those who hunt them sys-
tematically with decoys.

Only these exceptions and our summer resident species
will be mentioned here. Commonest among the latter
Woodcock,   is the Woodcock, a bird so unlike other
Philohela minor.   Snipe in his choice of haunts that he
Figs. 9 and 19.   seems quite out of place in this family.
Nor is he, strictly speaking, a summer resident, for there
are only three months in the year when the Woodcock
is not with us. He comes in March as soon as the frost-
bound earth will permit him to probe for his diet of
worms, and he remains until some December freeze
drives him southward.

Low, wet woods, where skunk cabbage and hellebore
thrive, or bush-grown, springy runs, are the Woodcock's
early haunts. In August, while molting, he often visits
cornfields in the bottom lands, and in the fall wooded
hillsides are his resorts. But, wherever he is, the Woodcock leaves his mark in the form of "borings"—little holes
which dot the earth in clusters, and show where the bird

has probed for earthworms with his long, sensitive bill, the upper mandible of which, as Mr. Gordon Trumbull has discovered, the bird can use as a finger.

The Woodcock's nest is made of dried leaves, and the four large, pear-shaped eggs are buff, spotted with shades of reddish brown. The young are born covered with rich chestnut and buff down, and can run as soon as dry.

As a songster the Woodcock is unique among our summer birds. Ordinarily sedate and dignified, even pompous in his demeanor, in the spring he falls a victim to the passion which is accountable for so many strange customs in the bird world.

If some April evening you visit the Woodcock's haunts at sunset, you may hear a loud, nasal note repeated at short intervals—*peent, peent.* It resembles the call of a Nighthawk, but is the Woodcock sounding the first notes of his love song. He is on the ground, and as you listen, the call ceases and the bird springs from the ground to mount skyward on whistling wings. He may rise three hundred feet, then, after a second's pause, one hears a twittering whistle and the bird shoots down steep inclines earthward. Unless disturbed, he will probably return to near the spot from which he started and at once resume his *peenting.* This, with the twittering note, is vocal; the whistling sound, heard as the bird rises, is produced by the rapid passage of air through its stiffened primaries.

Our only other common summer resident Snipe is the Spotted Sandpiper. It frequents the shores of lakes,

Spotted Sandpiper, ponds, and rivers, and is also found *Actitis macularia.* near the sea, but wherever seen may be Plate XI. known by its singular tipping, tetering motion, which has given it the names of Tip-up and Teter Snipe. It is also called Peet-weet, from its sharp call, rapidly repeated as it flies over the water. After gaining headway it sails for some distance, when its wide-stretched wings show a white bar or band.

The Spotted Sandpiper arrives from the South late in April and remains until October. It nests in the latter half of May, laying four pear-shaped eggs, in color white or buff, thickly spotted and speckled with chocolate, chiefly at the larger end. The young, like those of all Snipe, are born with a covering of downy feathers, and can run as soon as dry. The egg is, therefore, large in proportion to the size of the bird, and measures 1·25 by ·95 inches. (See Fig. 24*a*.)

Unlike the two preceding birds, Wilson's or the English Snipe is not a summer resident in the Middle

Wilson's Snipe,
Gallinago delicata.
Plate IX.

States, but as a rule nests from northern New England northward, though there are records of its breeding as far south as Connecticut and Pennsylvania. It migrates northward in March and April, and the return journey occurs during September and October. It is not a true shore bird, but frequents fresh-water marshes and meadows, and in rainy April weather, when the lowlands become more or less flooded, it may be found in places where few persons would think of looking for Snipe.

Like the Woodcock, Wilson's Snipe probes the mud for food, and when on the ground among the grasses its colors and pattern of coloration so closely resemble its surroundings that it is almost invisible.

When flushed, it utters a startled *scaip*, and darts quickly into the air, flying at first in so erratic a manner that it has become famous among sportsmen as a difficult mark.

Like the Nighthawk, Wilson's Snipe sometimes dives earthward from high in the air, making as he falls a sound which Minot compares to that produced by throw-

PLATE IX.

WILSON'S SNIPE.

Length, 11.25 inches. Upper parts black, buff, and rusty; throat and belly white, rest of under parts black and buff.

105

ing a nail held crosswise in the hand, though it is louder and more full. This performance is generally restricted to late evening and early morning during the spring, but is occasionally practiced in the fall.

Most of our transient visitant Snipe are true shore birds. Many of them are classed as game birds, and have

Semipalmated Sandpiper,
Ereunetes pusillus.
Plate X.

now become so uncommon that, as before remarked, it requires a special knowledge of their ways in order to find them. But there are some species too small to be worthy the sportsman's attention, and they are often numerous on our beaches. They are generally known as Peeps or Ox-eyes, but in books are termed Semipalmated Sandpipers — active little fellows, with black, gray and rusty backs and white under parts, who run along the shore, feeding on the small forms of life cast up by the waves. They are sociable birds, and even when feeding the members of a flock keep together, while when flying they move almost as one bird.

These Sandpipers visit us in May, when journeying to their summer homes within the Arctic Circle, and return in July, to linger on our shores until October. Their call-note is a cheery, peeping twitter, which probably suggested one of their common names.

PLOVERS. (FAMILY CHARADRIIDÆ.)

Most Plovers differ from Snipe in possessing three instead of four toes, and in having the scales on the tarsi rounded, not square or transverse. Their bill is shorter and stouter than that of Snipe, and they do not probe for food, but pick it up from the surface.

Although several species visit dry fields and uplands, they are ranked as shore birds or bay birds, and, as with Snipe, the species large enough to be ranked as game

Ernest Seton Thompson.

On the Shore

PLATE X.

COMMON TERN.
(*Length, 15·00 inches.*)

SEMIPALMATED SANDPIPER.   SEMIPALMATED PLOVER.
(*Length, 6·30 inches.*)   (*Length, 6·75 inches.*)

have become comparatively rare. Of the one hundred
known species, six visit eastern North America—the
Black-breasted, Golden, Piping, Wilson's, Semipalmated,
and Killdeer Plovers. Only the last two of these are
common enough to deserve mention here.

Killdeer,   The Killdeer, with the exception of
Ægialitis vocifera.   the Piping Plover, is the only bird of
Plate XI.   this family that nests with us. It is
irregularly distributed in the northeastern States, but its
noisy call, *kildee, kildee,* and striking markings render it
a conspicuous bird even where it is uncommon. It frequents uplands and lowlands, fields and shores, but prefers
the vicinity of water. Its nest of grasses is made on the
ground, and its four eggs are whitish, spotted and scrawled
with chestnut, chiefly at the larger end.

The Semipalmated or Ring-necked Plover looks like
a miniature of the Killdeer, but, in addition to other dif-
Semipalmated Plover,   ferences, has only one band on the
Ægialitis   breast. The male has the upper parts
    semipalmata. brownish gray, the under parts, nape,
Plate X.   and forehead white, while the breast-
band, crown, and cheeks are black. In the female these
black areas are brownish gray. This Plover visits our
shores and beaches during its northward migration in
May and southward migration in August and September.
Thanks to its small size, it is not hunted as game, and
for this reason is almost as common as the little Peeps
or Ox-eyes, with which it often associates. Its call is a
simple but exceedingly sweet and plaintive two-noted
whistle.

PLATE XI.

SPOTTED SANDPIPER.

Length, 7.50 inches. *Adult*, upper parts brownish gray and black ; under parts white spotted with black ; a white patch in wing. *Young*, similar, but without black.

KILLDEER.

Length, 10 50 inches   Upper parts brownish gray, upper tail-coverts rusty ; under parts white ; two bands on breast, crown and lores black, forehead and nape white.

GALLINACEOUS BIRDS. (ORDER GALLINÆ.)

BOB-WHITES, GROUSE, ETC. (FAMILY TETRAONIDÆ.)

THIS is the family of the game birds—the aristocrats of the bird world. They are protectively colored birds, their rich brown, buff, and black plumage harmonizing with their surroundings. Relying on their inconspicuousness, they avoid danger by hiding rather than by flight, taking wing only as a final resort. Then, with a startling *whir-r-r*, they spring into the air, their short, strong wings enabling them to reach their greatest speed within a short distance of the starting point.

One of the best-known members of this distinguished family is our familiar Bob-white, the Quail of the North and Partridge of the South. The fact is, he is neither a true Quail nor Partridge, and those who claim that but one of these names is correct may compromise on "Bob-white."

Bob-white,
Colinus virginianus.

The Bob-white inhabits the eastern United States, and wherever found is resident throughout the year. The sexes are much alike in color, the only important difference being in the throat and the line over the eye, which are white in the male and buff in the female.

No bird better illustrates the peculiar potency of bird song, and the hopelessness of attempting to express its charm. If I should describe Bob-white's call to a person who had never heard it, as two ringing notes, do you suppose he would have the faintest conception of what

110

they mean to those who love them? The promise of
Spring, its fulfillment in summer, is clearly told in Bob-
white's greeting. Then, in the autumn, when the members of a scattered bevy are signaling each other, their
sweet *where are you? where are you?* is equally associated
with the season.

The Bob-white nests about May 20, laying from ten
to eighteen white eggs in a nest on the ground.

The Ruffed Grouse, or Partridge of the North and
Pheasant of the South, is properly a true Grouse, and

Ruffed Grouse, can not be correctly called either Par-
Bonasa umbellus. tridge or Pheasant. He is a more
Plate XII. northern bird than the Bob-white, being found south of Virginia only in the Alleghanies.
Requiring large tracts of woodland for his haunts, he
is less generally distributed and not so common as his
plump relative.

I always associate the Grouse with the astounding
roar of wings made by the bird as he springs from the
ground at my feet and sails away through the forest. I
watch him at first with dazed surprise, then with a keen
sense of pleasure in the meeting. One need not be a
sportsman to appreciate the gaminess of the Grouse.

To find a hen Grouse with young is a memorable
experience. While the parent is giving us a lesson in
mother-love and bird intelligence, her downy chicks are
teaching us facts in protective coloration and heredity.
How the old one limps and flutters! She can barely
drag herself along the ground. But while we are watching her, what has become of the ten or a dozen little
yellow balls we almost stepped on? Not a feather do
we see, until, poking about in the leaves, we find one
little chap hiding here and another squatting there, all
perfectly still, and so like the leaves in color as to be
nearly invisible.

The drumming of the Grouse, as described by Mr. Thompson, begins "with the measured thump of the big drum, then gradually changes and dies away in the rumble of the kettle-drum. It may be briefly represented thus: *Thump—thump—thump—thump, thump ; thump, thump—rup rup rup rup, r-r-r-r-r-r-r-r-r.* The sound is produced by the male bird beating the air with his wings as he stands firmly braced on some favorite low perch."

The Ruffed Grouse makes its leaf-lined nest usually at the base of a tree or stump, and the eight to fourteen buff eggs are laid early in May.

PIGEONS AND DOVES. (ORDER COLUMBÆ.)

PIGEONS AND DOVES. (FAMILY COLUMBIDÆ.)

THE three hundred species belonging in this order are distributed throughout most parts of the world, but only two of them are found in the northeastern States. One of these, however, the Wild Pigeon, is now so rare that its occurrence is worthy of note. Less than fifty years ago it was exceedingly abundant, but its sociable habits of nesting and flying in enormous flocks made it easy prey for the market hunter, and, with that entire disregard of consequences which seems to characterize man's action when his greed is aroused, the birds were pursued so relentlessly that they have been practically exterminated.

The Mourning or Carolina Dove has happily been more fortunate. Nesting in isolated pairs, and not

Mourning Dove, gathering in very large flocks, it has
Zenaidura macroura. escaped the market hunter.

Plate XIII.   This Dove is found throughout the greater part of North America. In the latitude of New York it is a summer resident, arriving in March and

Ready to Drum

PLATE XII.

RUFFED GROUSE.

Length, 17 00 inches. *Male,* neck tufts long, black ; upper parts and tail gray or rusty, black and buff ; under parts white, black and rusty. *Female,* similar, but with neck tufts no longer than adjoining feathers.

113

9

remaining until November. In April we may hear its soft, sweet call, *coo-o-o, ah-coo-o-o—coo-o-o—coo-o-o,* as sad as the voice of the wind in the pines.

Although the bird is as beautiful in appearance as it is graceful in flight, it is a surprisingly poor housekeeper. Its platform nest of a few twigs is about as flimsy as any- thing worthy the name can be, and one wonders how even two eggs are kept on it long enough to hatch. In the West the nest is placed on the ground; in the East, on the lower branch of a tree.

Like all the members of their family, Doves immerse the bill while drinking, and do not withdraw it until the draught is finished. The young are fed on softened food regurgitated from the parent's crop.

BIRDS OF PREY. (ORDER RAPTORES.)

AMERICAN VULTURES. (FAMILY CATHARTIDÆ.)

THERE are but eight Vultures in the western hemi- sphere, and only two of these, the Black and the Turkey Vulture, are found in the eastern United States. The former is not often seen north of North **Turkey Vulture,** Carolina, but the Turkey Vulture, or *Cathartes aura.* Turkey Buzzard, as it is more frequent- ly called, comes each summer as far as Princeton, N. J., and occasionally strays farther north.

The Turkey Buzzard is one of Nature's scavengers, and, as such, is one of the few birds whose services to mankind are thoroughly appreciated. There are others of equal or even greater value who daily earn their right to the good will which we stupidly and persistently refuse to grant them; but of the Turkey Buzzard's assist- ance we have frequent convincing proof, and the decree has gone forth that injury to this bird is punishable by fine.

PLATE XIII.

MOURNING DOVE.

Length, 11·75 inches. Upper parts olive grayish brown, sides and back of neck iridescent ; breast with a pinkish tinge, belly buff ; outer tail-feathers tipped with white.

115

No other birds are so well protected ; and as a result Turkey Buzzards and Black Vultures walk about the streets of some of our Southern cities with the tameness of domestic fowls.　If we should similarly encourage our insectivorous birds, who can predict the benefits which might accrue ?

HAWKS, FALCONS, AND EAGLES. (FAMILY FALCONIDÆ.)

To this family belong the diurnal birds of prey, which number some three hundred and fifty species, and are distributed throughout the world.　They are birds of strong flight, and capture their prey on the wing by striking it with their sharp, curved claws, the most deadly weapons to be found in any bird's armament.　The bill is short, stout, and hooked, and is used to tear the prey while it is held by the feet.

The voices of Hawks are in keeping with their dispositions, and, while their lives typify all that is fierce and cruel, no birds are more often wrongly accused and falsely persecuted than our birds of prey.　To kill one is regarded as an act of special merit; to spare one seems to place a premium on crime.　Still, these birds are among the best friends of the farmer.　There are but two of our common species, Cooper's and the Sharp-shinned, who habitually feed on birds and poultry.　Our other common species are, without exception, invaluable aids to the agriculturist in preventing the undue increase of the small rodents so destructive to crops.

Any one reading Dr. Fisher's reports on this subject can not fail to be impressed with the array of facts he presents in proof of the value of these

Red-shouldered Hawk,
Buteo lineatus.
Plate XIV.

birds.　For instance, the Red-shouldered Hawk, to which the name Chicken or Hen Hawk is often applied, has been found to live largely on small mammals,

Plate XIV.

RED-SHOULDERED HAWK.

Length, 19 00 inches. *Adult,* upper parts blackish brown and rusty ; lesser
wing-coverts bright chestnut : wings and tail black and white ; under parts
rich rusty and white. *Young,* less rusty on back, wings and tail largely
rusty ; under parts white, spotted or streaked with blackish.

117

reptiles, batrachians, and insects. Indeed of 220 stomachs which were examined of this so-called " Chicken " Hawk, only 3 contained remains of poultry! Of the rest, 12 contained birds; 102, mice; 40, other mammals; 20, reptiles; 39, batrachians; 92, insects; 16, spiders; 7, crawfish; 1, earthworms; 2, offal; 3, fish; and 14 were empty.* The usefulness of this Hawk is therefore obvious, and in killing it we can readily see that we not only harm ourselves but render an important service to our enemies.

Fortunately, this valuable ally is one of our commonest Hawks, and is with us throughout the year. Its loud scream, *kèe-you, kèe-you,* as it sails about, high in the air, is a familiar summer sound. The " red " shoulder is in reality a rich, reddish chestnut on the lesser wing-coverts, and serves to identify the bird in both immature and adult plumage. The Red-shoulder's nest, like that of most of our Hawks, is constructed of sticks and twigs, with a lining of cedar bark, moss, or some other soft material, and is situated in a tree thirty to sixty feet from the ground. Apparently the same pair of birds return to a locality year after year, sometimes using the same nest, at others building a new one. The eggs are about as large as those of a hen and in color are dull white, more or less sprinkled, spotted, or blotched with cinnamon-brown or chocolate. They are laid early in April, most of the Hawks being early breeders. The young are born covered with white down, but are helpless, and are reared in the nest.

The Red-tailed Hawk is also known as the Hen Hawk or Chicken Hawk, but has almost as good a record as

---

* See Fisher, The Hawks and Owls of the United States in their Relation to Agriculture; Bulletin No. 3, Division of Ornithology and Mammalogy, United States Department of Agriculture, 1893.

PLATE XV.

MARSH HAWK.

Length, 20 00 inches. *Adult male,* upper parts gray ; under parts white with
rusty spots ; upper tail-coverts white. *Adult female and young,* upper
parts black and rich rusty ; under parts rich rusty and black ; upper tail-
coverts white.

the Red-shoulder, and is equally deserving of protection. He is larger than the Red-shoulder, whom he resembles in habits, and has a reddish

Red-tailed Hawk,
Buteo borealis.

brown tail and a broken black band across the breast when adult. His call is a thin, long-drawn, wheezy whistle, which reminds one of the sound produced by escaping steam.

The Marsh Hawk courses to and fro over field and meadow, like a Gull over the water. He never sails,

Marsh Hawk,
Circus hudsonius.
Plate X V.

however, but on firm wing flies easily and gracefully, ever on the watch for prey in the grasses below. He may sometimes mistake birds for mice, but he captures far more of the latter than of the former, and only 7 of the 124 Marsh Hawks whose stomachs were examined by Dr. Fisher had eaten chickens.

The Marsh Hawk is migratory, and in winter is not often found north of southern Connecticut. He nests later than the resident Hawks, and, unlike them, builds his nest of grasses on the ground in the marshes, laying from four to six dull white or bluish white eggs early in May.

The Sparrow Hawk has a perfectly clean record, as far as chickens go, not one of the 320 whose stomachs

Sparrow Hawk,
Falco sparverius.
Plate XVI.

were examined by Dr. Fisher, having partaken of poultry, while no less than 215 had eaten insects, and 89 had captured mice. Grasshoppers are the Sparrow Hawk's chief food, and we may often see him hovering over the fields with rapidly moving wings. Then, dropping lightly down on some unsuspected victim below, he returns to the bare limb or stub he uses for a lookout station, uttering an exultant *killy—killy—killy* as he flies.

The Sparrow is distributed throughout the greater part of North America, but in winter is not found north

Plate XVI.

SPARROW HAWK.

Length, 11·00 inches   *Male*, back reddish brown and black, wing-coverts slaty blue, tail reddish brown marked with black and white ; under parts washed with rusty and spotted with black.   *Female*, back, wings, and tail barred with reddish brown and black ; under parts white, streaked with reddish brown, of southern New York. It migrates northward in February and March, but does not nest until May. Unlike our other Hawks, it chooses a hollow tree for a home, often taking possession of a Woodpecker's deserted hole. It lays three to seven eggs, which are finely and evenly marked with reddish brown.

It is the Sharp-shinned and Cooper's Hawks who are the real culprits in Hawkdom. They feed almost exclu-

Sharp-shinned Hawk,
Accipiter velox.
Plate XVII.

sively on birds, and, having once acquired a taste for tender young broilers, they are apt to make daily visits to the hen yards. They are less often observed than the Hawks previously mentioned, seeking less exposed perches and soaring comparatively little ; but, when seen, their slender bodies and long tails should aid in distinguishing them from the stouter, slower-flying Hawks. As a rule, they are silent. It is difficult to explain the differences between these and other Hawks with sufficient clearness to prevent one's killing the wrong kind, but if the farmer will withhold his judgment against Hawks in general, and shoot only those that visit his poultry yard, he will not go far astray.

Cooper's Hawk resembles the Sharp-shinned in color, but is about four inches longer, and its outer tail-feathers

Cooper's Hawk,
Accipiter cooperi.

are about half an inch shorter than the middle ones instead of being of equal length. With the preceding species it may be known by its slender form, long tail, comparatively short wings, and long, thin tarsi or "legs."

The Chinese and Japanese train Cormorants to fish

American Osprey,
Pandion haliaetus carolinensis.
Plate XVIII.

for them, but the services of these birds would soon be at a discount if the Osprey could be induced to work for a master. What an inspiring sight it is to see one plunge from the air upon its prey ! One can

SHARP-SHINNED HAWK.

Length of male, 11·25 inches; of female, 13·50 inches. *Adult*, upper parts slaty gray; under parts white and rusty brown. *Young*, upper parts blackish brown; under parts white, streaked with rusty brown.

sometimes hear the splash half a mile or more, and the bird is quite concealed by the spray. It is a magnificent performance, and when, after shaking the water from his plumage, he rises into the air, I am always tempted to applaud.

The Osprey, or Fish Hawk, as he is also called, adheres closely to a finny diet; neither flesh nor fowl appears on his *menu*, and he is consequently a migratory bird, coming in April when the ice has melted and remaining until October. In favorable localities he nests in colonies, returning year after year to the same nest.

One master, it is true, the Osprey has, though he makes a most unwilling servant. The Bald-headed Eagle is often an appreciative observer of the Osprey's

Bald Eagle, piscatorial powers, which so far exceed
Haliæetus his own that he wisely, if unjustly,
leucocephalus. profits by them. Pursuing the Osprey,

he forces him to mount higher and higher until the poor bird in despair drops his prize, which the Eagle captures as it falls.

Eagles are becoming so rare in the Northern States that their occurrence is sometimes commented on by the local press as a matter of general interest. Nevertheless, no opportunity to kill them is neglected, and the majestic birds who in life arouse our keenest admiration are sacrificed to the wanton desire to kill.

THE OWLS. (FAMILY BUBONIDÆ.)

The Owls number about two hundred species, and are distributed throughout the world. As a rule they are nocturnal or crepuscular birds, passing the day in hollow trees or dense evergreens, and appearing only after nightfall; but there are some diurnal species, such

PLATE XVIII.

AMERICAN OSPREY.

Length, 23·00 inches. Upper parts brownish black ; nape and under parts white ; breast marked with grayish brown.

125

as the Snowy Owl and Hawk Owl, northern birds that
visit us rarely in winter.

Because of their nocturnal habits Owls are even more
deserving of protection than the beneficial Hawks, for
they feed at a time when mice are abroad, and their
food consists largely of these destructive little rodents.
They capture their prey, like the Hawks, by striking it
with their powerful talons, when, if small enough, it is
swallowed entire. The indigestible portions, hair, bones,
and feathers, are formed into pellets in the stomach and
ejected at the mouth. These may always be found in
numbers beneath an Owl's roosting place, and form as
sure an indication of the Owl's presence as they do of
the nature of his food. Thus, as before mentioned, two
hundred pellets of the Barn Owl, taken from the home of
a pair of these birds in the tower of the Smithsonian In-
stitution, were found by Dr. A. K. Fisher to contain the
skulls of 454 small mammals.

Owls are generally inhabitants of woods, but our
Short-eared Owl is an exception to this rule, and lives

Short-eared Owl, in large, grassy marshes. It passes the
Asio accipitrinus. day on the ground, but at dusk may be
Plate XIX. seen flying low over the marsh in search
of the meadow mice which form a large part of its food.
Dr. Fisher found, on examination of 101 stomachs of this
Owl, that no less than 77 contained the remains of mice,
convincing proof of its usefulness. Unlike any of our
other Owls, the Short-eared makes its nest on the ground,
laying from four to seven eggs. It is somewhat irregular
in its distribution, but has been found nesting, locally,
from Virginia northward. It winters from New Jersey
southward, and is sometimes associated in companies at
this season.

The Long-eared Owl is about the size of the Short-
eared Owl, but its "ear-tufts" are an inch or more in

SHORT-EARED OWL.

Length, 15·50 inches.   Upper parts black, buff, and rusty ;   under parts white
and brownish black ;   eyes yellow.

length, and its sides and belly are *barred*, not *streaked*, with blackish. It does not frequent marshes, but lives in swampy thickets or dense woods, and

Long-eared Owl,
Asio wilsonianus.

makes its nest in the abandoned home of a Crow, Hawk, or squirrel. It is a permanent resident from at least Massachusetts southward.

Of our four "horned" Owls, the Long-eared has relatively the largest and most conspicuous "ear-tufts," the Short-eared the smallest, while in the Great Horned Owl and Screech Owl the ears are of about the same proportionate size. The Great Horned Owl, however, is found only in the wilder, more heavily wooded parts of the country, and is hardly to be included in a list of our common birds. It is the largest of our resident Owls, the males measuring twenty-two inches in length, while its "ear-tufts" are nearly two inches long.

The Screech Owl is doubtless the commonest of our Owls, as it is also the most familiar, nesting about and

Screech Owl,
Megascops asio.
Plate XX.

even in our houses when some favorable hole offers. It has little to say for itself until its family of four to six fuzzy Owlets is safely launched into the world; then, in July or August, we may hear its melancholy voice—not a "screech," but a tremulous, wailing whistle. It has several other notes difficult to describe, and when alarmed defiantly snaps its bill.

Some Screech Owls are gray, others bright reddish brown, and these extremes are connected by specimens intermediate in color. This difference in color is not due to age, sex, or season, and is termed dichromatism, or the presence in the same species of two phases of color. The same phenomenon is shown by other birds, notably certain Herons, and among mammals by the gray squirrel, some individuals of which are black. The observa-

PLATE XX.

SCREECH OWL.

Length, 9·40 inches.   Upper parts gray, or bright reddish brown, and black ;
under parts white, gray, or bright reddish brown, and black ; eyes yellow.

149

10

tions of Dr. A. P. Chadbourne apparently show that the Screech Owl may pass from one phase to another without change of plumage.*

We do not think of Owls as being insectivorous birds, but Dr. A. K. Fisher tells us that of 225 Screech Owls' stomachs examined, 100 contained insects. As 91 of the remaining 125 contained mice, and poultry was found in only one stomach, the farmer may well consider the Screech Owl a bird of good repute rather than of ill omen.

Next to the Screech Owl the Barred Owl is doubtless our most common representative of this family, but its **Barred Owl,** fondness for deep woods prevents its *Syrnium nebulosum.* being known to many who recognize the Plate XXI. Screech Owl's mournful song.

In both voice and appearance the Barred Owl seems the most human of our Owls. Its call is a deep-voiced questioning *whŏŏ-whŏŏ-whŏŏ, whŏ-whŏŏ, tŏ-whōō-ah,* which may be heard at a distance of half a mile. It echoes through the woods at night with startling force, and the stories told of its effect on persons who were ignorant of its source are doubtless not without foundation.

Other calls are a long-drawn *whō-ō-ō-ō-āh,* and rarely a thrilling, weird shriek. When two or more Owls are together, they sometimes join in a most singular concerted performance. One utters about ten rapid hoots, while the other, in a slightly higher tone, hoots about half as fast, both birds ending together with a *whōō-ah.* At other times they may *hoot* and laugh in a most remarkable and quite indescribable manner.

The Barred Owl feeds largely on mice, and 46 of 100 stomachs examined contained remains of these rodents.

---

* The Auk (New York city), xiii, 1896, p. 321 ; xiv, 1897, p. 33.

BARRED OWL.

Length, 20·00 inches. Upper parts blackish brown and white; under parts white and blackish brown; eyes black.

It is generally resident throughout its range, and in March makes its nest, selecting for a site a hollow tree, or the deserted home of a Crow or Hawk. Two to four eggs are laid, which, like the eggs of all Owls, are pure white.

CUCKOOS, KINGFISHERS, ETC. (ORDER COCCYGES).

CUCKOOS. (FAMILY CUCULIDÆ.)

ALL Cuckoos have two toes directed forward and two backward, but the cause or use of this character it is difficult to understand, so widely do the members of this family differ in habit. Some are arboreal, never visiting the earth, while others are terrestrial, running with great swiftness, and rarely perching far above the ground.

Most Cuckoos—all our thirty-five American species— have noticeably long tails, which they raise and droop slowly just after alighting, or when their curiosity is aroused.

Of the one hundred and seventy-five known species, only two are found in the northeastern States—the Yel-

Yellow-billed Cuckoo,
Coccyzus americanus.
Plate XXII.

billed and the Black-billed Cuckoos. The former is generally the more common. It is a retiring bird, and you will doubtless be first attracted to it by its notes. It does not perch in an exposed position, nor make long flights, but usually flies from the shelter of one tree directly into the protecting foliage of another. If you catch a glimpse of it in passing, its long tail and brownish color will suggest a Dove.

Cuckoos are mysterious birds well worth watching. I would not imply that their deeds are evil; on the contrary, they are exceedingly beneficial birds. One of their favorite foods is the tent caterpillar which spins the

PLATE XXII.

YELLOW-BILLED CUCKOO.

Length, 12 25 inches. Upper parts glossy olive-brown ; outer tail-feathers
black, tipped with white ; under parts white ; lower mandible yellow.

destructive "worms' nests" in our fruit and shade trees. Indeed, we should be very much better off if Cuckoos were more numerous. Nevertheless, there is something about the Cuckoo's actions which always suggests to me that he either has just done, or is about to do, something he shouldn't.

The Yellow-billed Cuckoo's call begins with a series of *tut-tuts* or *cl-ucks*, and ends with a loud *cow, cow, cow, cow, cow, cow*. These notes are so unlike those of any other of our birds, except those of the Black-billed Cuckoo, that they will readily be recognized.

The Black-billed Cuckoo resembles the Yellow-bill, but has the bill wholly black, the skin about the eye red, and the tail-feathers with only small, inconspicuous whitish tips. It resembles the Yellow-bill in habits, but, as Mr. Brewster tells me, its *tut* and *cluck* notes are softer, and the *cow-cow* notes are connected.

Black-billed Cuckoo,
Coccyzus
erythrophthalmus.

Both our Cuckoos are migratory, wintering in Central and South America. They return to us about May 5, and remain until October. Their nests are carelessly made platforms of sticks with a few catkins added as a lining, and are usually placed in tangles of vine-covered bushes, or the lower limbs of trees. The eggs, three to five in number, are pale, greenish blue, those of the Black-bill being slightly smaller in size and darker in color than those of its yellow-billed cousin.

KINGFISHERS. (FAMILY ALCEDINIDÆ.)

Of the one hundred and eighty known Kingfishers, only eight are inhabitants of the New World, the headquarters of the family being in the East Indies. The New World species are mostly tropical, and but one of the eight reaches the eastern United States. This is our common

PLATE XXIII.

BELTED KINGFISHER.

Length, 13·00 inches. *Male*, upper parts bluish gray ; under parts white, a
bluish gray breast-band and sides. *Female*, similar, but breast and sides
with reddish brown.

Belted Kingfisher, familiar by voice and appearance to
every one who lives near a river or pond. He comes

Belted Kingfisher, in April, when the ice no longer cov-
Ceryle alcyon. ers his hunting ground, and remains
Plate XXIII. until November; or, if the season be
exceptionally mild, he sometimes stays for the winter
fishing. His nest is built in a hole in a bank, where,
early in May, his-mate lays from five to eight white
eggs.

The Kingfisher is generally branded a fish thief and
accounted a fair mark for every man with a gun, and,
were it not for his discretion in judging distances and
knowing just when to fly, he would long ago have disappeared from the haunts of man. We might now be a
few fish richer, but would they repay us for the loss of
this genius of wooded shores?

WOODPECKERS AND WRYNECKS. (ORDER PICI.)

WOODPECKERS. (FAMILY PICIDÆ.)

THE three hundred and fifty known species of Woodpeckers are represented in all the wooded parts of the
world except the Australian region and Madagascar.
Nearly one half this number are found in the New
World, and of these twenty-five occur in North America.

Few birds seem better adapted to their mode of life
than Woodpeckers, the structure of their bill, tongue,
tail, and feet being admirably suited to their needs.

The notes of Woodpeckers can not be termed musical,
and their chief contribution to the springtime chorus is a
rolling tattoo which resembles the *k-r-r-r-ring* call of the
tree frogs. The feathered drummer selects a resonant
limb and pounds out his song with a series of strokes de-

A Woodland Carpenter.

DOWNY WOODPECKER.

Length, 6·75 inches.　*Male*, upper parts black and white, nape scarlet ; under parts white.　*Female*, similar, but no scarlet on nape.

livered so quickly that his head becomes a series of mazy heads.

Watch the Downy Woodpecker, our commonest species, while he is engaged in this surprising perform-

Downy Woodpecker,
Dryobates pubescens medianus.
Plate XXIV.

ance. How he seems to enjoy it! His whole appearance is martial and defiant. It is his challenge to the Woodpecker world. After each roll he looks proudly about him and perhaps utters his call-note, a sharp *peek, peek,* which suggests the sound produced by a marble cutter's chisel. More rarely this call is prolonged into a connected series, when one can readily imagine that the quarrier has dropped his tool.

The Downy is a hardy bird and is with us throughout the year. In the winter he forms a partnership with the Chickadee and Nuthatch, and if the good this trio does could be expressed in figures, these neglected friends of ours might receive some small part of the credit due them. Who can estimate the enormous numbers of insects' eggs and larvæ which these patient explorers of trunk and twig destroy?

The Downy, as well as some other Woodpeckers, believes in the comfort of a home. He will not pass cold, wintry nights clinging to the leeward side of a tree when by the use of his chisel-bill he can hollow a snug chamber in its heart. So, in the fall, we may sometimes find him preparing his winter quarters. His nest is constructed in the same manner, and his eggs, like those of all Woodpeckers, are glossy white.

The Hairy Woodpecker, the Downy's big cousin, is

Hairy Woodpecker,
Dryobates villosus.

not quite so common as his smaller relative. The two birds are nearly alike in color, and differ only in the markings of the outer tail-feathers. In the Downy these are white, barred with black; in the Hairy, white without

RED-HEADED WOODPECKER.

Length, 9·75 inches. *Adult*, whole head and neck deep red, back and tail black ; upper tail-coverts, greater part of secondaries, and belly white. *Young*, similar, but head, back, throat, and sides grayish black.

black bars. The case is interesting, and shows how nearly alike in color distinct species may be. In size, however, the difference is more noticeable, the Hairy being nearly three inches the longer.

In life the Hairy is a somewhat shier bird, fonder of the forest than of the orchard. His *peek* note is louder and sharper than that of the Downy, and his rattling call suggests that of the Kingfisher.

The gayly colored Red-headed Woodpecker is as erratic in his goings and comings as he is striking in dress. In the northeastern States he is **Red-headed Woodpecker.** locally common in summer, and if well *Melanerpes erythrocephalus.* supplied with beechnuts, may remain Plate XXV. during the winter. Some years the grayish headed young birds are exceptionally abundant in the fall, but their white wing-patches, which show so conspicuously when they fly, and their loud, rolling call of *ker-r-ruck, ker-r-ruck,* are unmistakable marks of identity.

The most interesting of our Woodpeckers is the Flicker, or High-hole, whose popularity is attested by **Flicker,** his list of no less than thirty odd com-*Colaptes auratus.* mon names. Surely here is an instance Plate XXVI. illustrating the necessity of one scientific term by which the " Piquebois jaune " of Louisiana may be recognized as the " Clape " of New York. He is also a Yucker, a Flicker, and a Yellow-hammer; all these names being based on his notes or plumage.

The Flicker is less of a carpenter than are others of his family, and generally selects decayed logs and stumps as his hunting grounds. Here he hunts for his favorite food of ants, which he also procures at their holes and mounds. This is the reason we so often flush the Flicker from the ground, and, if we mark the spot from which he

PLATE XXVI.

FLICKER.

Length, 12·00 inches. *Male*, crown gray, nape scarlet, back brownish and black, rump white ; under surface of wings and tail yellow : sides of throat and breast-patch black ; belly spotted with black. *Female*, similar, but no black on sides of throat.

rises, the probabilities are that we shall find there a much-disturbed community of ants.

Professor Beal has shown that nearly one half of the Flicker's food consists of ants. He further tells us that as ants aid in the increase of the plant lice so injurious to vegetation, the birds which feed on ants are therefore the friends of the agriculturist.

The Flicker's most prominent marks, as with a low chuckle he bounds up before you, are his white rump patch and his wings, which show yellow in flight. His notes are equally characteristic. The most common is a loud, vigorous *kèe-yer*, apparently a signal or salute. In the spring, and occasionally in the fall, he utters a pleasing, rather dreamy *cŭh-cŭh-cŭh-cŭh*, many times repeated. When two or more birds are together, and in my experience only then, they address each other with a singular *weèchew, weèchew, weèchew*, a sound which can be imitated by the swishing of a willow wand. Much ceremony evidently prevails in the Flicker family, and on these occasions there is more bowing and scraping than one often sees outside of Spain.

GOATSUCKERS, SWIFTS, AND HUMMINGBIRDS.
(ORDER MACROCHIRES.)

NIGHTHAWKS AND WHIP-POOR-WILLS.
(FAMILY CAPRIMULGIDÆ.)

IN this family the mouth of birds reaches its greatest development, while the bill proper is correspondingly small, bearing much the same relation to the mouth that a clasp does to a purse. These birds feed at night upon insects which they catch on the wing, and their enormous gape is obviously of great assistance in this mode of feeding. Often the sides of the mouth are beset with long

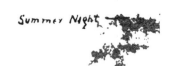

PLATE XXVII.

NIGHTHAWK.

Length, 10·00 inches. *Male*, above, black, white, and rusty ; below, black and white ; throat, bands in wing, and tail white. *Female*, similar, but throat rusty ; no tail-band.

WHIP–POOR–WILL.

Length, 9·75 inches. *Male*, body black, rusty, and buff ; primaries spotted with rusty ; tips of outer tail-feathers and breast-band white. *Female*, similar, but breast-band and end of tail rusty.

143

bristles, which doubtless act like the wings to a fish-net, steering unfortunate insects down the bird's cavernous throat.

The Nighthawk, or Bull-bat, as he is called in the South, is familiar to most persons who have the gift of seeing birds, but—in the northeastern States, at least—he is usually confused with the Whip-poor-will, and little is known of his real character.

Nighthawk,
Chordeiles
virginianus.
Plate XXVII.

The Nighthawk is a bird of the sky. He passes the day perched motionless on a limb in wooded regions, on the ground in treeless regions, or even on a house top, when, as sometimes happens, he makes his home in a city. Probably he will not change his perch during the day, but as night approaches and his day begins, he will spread his long wings and fly away heavenward to course far above the earth in his search for insect food.

The Nighthawk, unlike most members of its family, has limited vocal powers, its only note being a loud, nasal *peent* uttered as it flies. But it has musical talents in another direction. Sometimes in May or June, if you happen to be where Nighthawks are found—for they are rather local in distribution when nesting—you may hear a strange booming, rushing sound; you will vainly seek its cause until you chance to see a Nighthawk with set wings diving earthward from the sky. It is a reckless performance, and you may suppose the bird's object is suicidal, but, when within a few yards of the earth, it will turn suddenly upward. At this moment you will hear the loud, humming sound, doubtless made by the air passing through the bird's stiffened wing-quills.

Nighthawks, being insect-catchers, are of course highly migratory. They come to us early in May, and return to their winter quarters in South America in Oc-

PLATE XXVIII.

CHIMNEY SWIFT.

Length, 5·40 inches.   Sooty black, throat grayish.

11

tober. During the fall migrations they often gather in flocks of several hundred, and as they sail about you may notice their best field mark, a white spot in each wing. Nighthawks lay two elliptical, mottled eggs on the bare ground or a flat rock in open fields, and, rarely, on a house top in the city.

We see the Nighthawk and hear the Whip-poor-will; one reason perhaps why the birds are so often confused.

Whip-poor-will,
Antrostomus vociferus.
Plate XXVII.

While the Nighthawk is darting through the sky, the Whip-poor-will is perched on a rock or fence rail below, industriously whipping out a succession of rapid *whip-poor-wills* interspersed with barely audible *chucks*. When the call ceases, the bird is doubtless coursing low through the wooded fields and glades in its search for insects.

During the day the Whip-poor-will usually rests on the ground in the woods. Here also the eggs are laid, being deposited upon the leaves. They are two in number, dull white, with delicate, obscure lilac markings and a few distinct brownish gray spots.

Whip-poor-wills arrive from the south late in April, and remain with us until October.

SWIFTS. (FAMILY MICROPODIDÆ.)

SWIFTS are the most aërial of all the small land birds. Our Chimney Swift, the only one of the seventy-five

Chimney Swift,
Chætura pelagica.
Plate XXVIII.

members of this family that occurs in eastern North America, is but five and a half inches long, while its spread wings measure twelve and a half inches from tip to tip. Its feet are proportionately small, and so weak that the bird can rest only by clinging to an upright surface. The tail is then used as a prop, its spiny-tipped feathers being evidently a result of this habit.

PLATE XXIX.

RUBY-THROATED HUMMINGBIRD.

Length, 3.75 inches. *Adult male*, upper parts metallic green ; throat metallic ruby-red ; belly grayish ; sides greenish. *Adult female and young*, similar, but throat white.

147

Swifts naturally nest in hollow trees or caves, and it is only in the more densely populated parts of their range that they resort to chimneys and outbuildings. The nest of our Chimney Swift is a bracketlike basket of small twigs. They are gathered by the bird while on the wing, and are fastened together and to the wall of the tree or chimney with a glutinous saliva.

The Chimney Swift arrives from the south about April 20, and remains until October. Few birds are better known, and under the name of "Chimney Swallow" he is familiar to every one who distinguishes a Crow from a Robin. But, beyond similar feeding habits, Swifts have little in common with Swallows; in fact, are more nearly related to Hummingbirds.

HUMMINGBIRDS. (FAMILY TROCHILIDÆ.)

HUMMINGBIRDS are peculiar to the New World. About five hundred species are known, but only one of them is found east of the Mississippi. This is our Ruby-throat, the sexes of which are sometimes thought to represent different species. The Ruby-throat winters

Ruby-throated Hummingbird,
Trochilus colubris.
Plate XXIX.

as far south as Central America, but about May 1 we may expect him to return to us, for he is as regular in his migrations as though his wings measured a foot and a half instead of an inch and a half in length. If you would have him visit you, plant honeysuckle and trumpet flowers about your piazza, and while they are blooming there will be few days when you may not hear the humming of this tiny bird's rapidly vibrating wings.

The Ruby-throat feeds on insects as well as on the juices of flowers, and when you see him probing a corolla he is quite as likely to be after the one as the other. The young are fed by regurgitation, the parent bird inserting its bill into the mouth of its offspring and injecting food as though from a syringe.

Some tropical Hummingbirds have songs worthy the name, but the notes of our Ruby-throat are a mere squeak, sometimes prolonged into a twitter.

Under any circumstances a Hummingbird's nest excites admiration. But if you would appreciate its fairylike beauty, find one where the birds have placed it, probably on the horizontal limb of a birch. Doubtless it will be occupied by the female, for it seems that the male takes little or no part in family affairs after incubation begins. As far as known, all Hummingbirds lay two white eggs —frail, pearly ellipses, that after ten days' incubation develop into a tangle of tiny dark limbs and bodies, which no one would think of calling birds, much less " winged gems."

PERCHING BIRDS. (ORDER PASSERES.)

FLYCATCHERS. (FAMILY TYRANNIDÆ.)

DOUBTLESS, every order of birds has had its day when, if it was not a dominant type, it was at least sufficiently near it to be considered modern ; and as we review what is known to us of that great series of feathered forms, from the Archæopteryx to the Thrushes, we can realize how varied has been the characteristic *avifauna* of each succeeding epoch from the Jurassic period to the present.

Now has come the day of the order *Passeres*, the Perching Birds; here belong our Flycatchers, Orioles, Jays, Sparrows and Finches, Vireos, Swallows, Warblers, Wrens, Thrushes, and many others. A recent authority classifies birds in thirty-four orders, but fully one half of the thirteen thousand known species are included in the single order *Passeres*. The North American members of this order are so alike in more important structural details that they are placed in but two suborders, the suborder *Clamatores*, containing the so-called Songless Perching Birds, and the suborder *Oscines*, containing the Song Birds. The Flycatchers are the only members of the suborder *Clamatores* in Eastern North America. They differ from the *Oscines*, or true Song Birds, in always having ten fully developed primaries, in having the tarsus rounded behind as well as in front, and chiefly in the anatomy of the syrinx, or voice-producing organ. In the *Oscines* this possesses four or five distinct pairs of intrinsic muscles, while in the *Clamatores* it has less than four pairs of muscles, and is not so highly developed.

Flycatchers are the Hawks of the insect world. Their position when resting is erect, and they are constantly on the watch for their prey, which is captured on the wing, with a dexterity Hawks may well envy. The bill is broad and flat and the gape large, as in other fly-catching birds. After darting for an insect, as a rule, they return to the same perch, a habit which betrays their family affinities, though it is occasionally practiced by some other birds.

Among our Eastern Flycatchers the Kingbird undoubtedly deserves first rank. In books he is sometimes

Kingbird, called the Tyrant, but the name is a
Tyrannus tyrannus. libel. The Kingbird is a fighter, but
Plate XXX. he is not a bully, and gives battle only
in a just cause. His particular enemy is the Crow, and during the nesting season each Kingbird evidently draws an imaginary circle about his home within which no Crow can venture unchallenged. From his lookout on the topmost branch of a neighboring tree the Kingbird darts forth at the trespasser, charging him with a spirit

PLATE XXX.

KINGBIRD.

Length, 8·50 inches. Upper parts grayish black ; tip of tail and under parts
white ; an orange-red crown-patch. *Young,* similar, but without orange-red
in crown.

and fearlessness which no bird can withstand. It is a case of " right makes might," added to a very dexterous use of wings and bill. The Crow, if he be experienced, turns tail at once and, beyond protesting *squawks*, makes no attempt to defend himself. But the Kingbird is deaf to pleas for mercy; he too has had experience, and well knows that only his own watchfulness has saved his eggs or young. Far in the distance he relentlessly pursues his foe, leaving him only when he has administered a lesson which will not be forgotten. Then he returns to his post and, with crest erect and quivering wings, gives voice to cries of victory.

Bee-keepers accuse the Kingbird of a taste for honey-bees, but the examination, made by Prof. Beal, of two hundred and eighteen Kingbirds' stomachs shows that the charge is unfounded. Only fourteen stomachs contained remains of bees, most of which were drones, while sixty per cent of the Kingbirds' food was found to consist of injurious insects.

Kingbirds winter in Central and South America, returning to us in the spring about May 1, and remaining until September. Their nest is a compact, symmetrical structure of weed stalks, grasses, and moss, lined with plant down, fine grasses, and rootlets, and is usually placed at the extremity of a limb about twenty feet from the ground. The eggs, three to five in number, are white, spotted with chocolate.

The Crested or Great Crested Flycatcher is, as a rule, not so common as the Kingbird, and its habits prevent it

Crested Flycatcher,
Myiarchus crinitus.
Plate XXXI.

from being so easily observed. Kingbirds can be seen whenever heard, but you may hear the Greatcrest's whistle many times before you see the whistler. Generally he lives in the woods high up in the trees, but he is also found in old orchards. His call, like an exclamation,

CRESTED FLYCATCHER.

Length, 9 00 inches. Upper parts brownish olive-green ; inner vane of tail-feathers rusty ; breast gray ; belly pale yellow.

rings out above all other birds' notes. *What!* he seems to say, and, as though hearing something which not only surprised but amused him, follows this call with a chuckling whistle.

The Greatcrest arrives from the south about May 7, and remains until September. Nesting is begun early in June, a hollow limb being the home usually selected. In collecting its nesting materials, the bird displays a very singular trait, and gives evidence of the stability of habit. With rare exceptions it places a bit of cast snake-skin in its nest. Various reasons have been advanced to account for this singular habit, but none of them is satisfactory. Recently Lieutenant Wirt Robinson has discovered that one of the commonest and most generally distributed species of this genus in South America places cast snake-skin in its nest, and it is well known that the Arizona Crested Flycatcher follows the same custom. The habit is therefore widespread, and is common to birds living under greatly varying conditions. Rather than consider it of especial significance in each species, it seems more reasonable to believe that it is an inheritance from a common ancestor, and has no connection with the present surroundings of at least those species living so far from the center of distribution of this tropical genus as our *Myiarchus crinitus*.

The Phœbe is domestic; he prefers the haunts, or, at least, handiwork of man, and when not nesting on a beam in a barn, shed, or piazza, selects the **Phœbe,** *Sayornis phœbe.* Plate XXXII. shelter of a bridge for a home. Here he places his nest of moss and mud; a structure of generous proportions, for the Phœbe's family may number five or six.

Flycatchers, because of the nature of their food, usually make extended migrations. For the same reason they arrive late in the spring and depart early in the

PLATE XXXII.

PHŒBE.

Length, 7·00 inches. Back dusky olive ; crown blackish ; under parts white tinged with yellow ; outer margin of outer tail-feathers whitish ; bill black.

155

fall; but the Phœbe is an exception to this rule. Not only does he winter north of the frost line, but he comes to us as early as March 20 and remains until October.

The Phœbe owes his name to his song of *pewit-phœbe*, *pewit-phœbe*, a humble lay uttered between vigorous wags of the tail. This tail-wagging is a characteristic motion, and also accompanies the Phœbe's call-note, *pee, pee*, which it utters at intervals.

The Least Flycatcher shares the Phœbe's preference for the vicinity of houses and is most often found nesting in our shade or fruit trees. The nest,
Least Flycatcher,
Empidonax minimus. unlike the Phœbe's, is composed of plant-down, fibers, and rootlets, and is placed in the crotch of a tree. The eggs resemble the Phœbe's in being white.

It is difficult to describe our smaller Flycatchers so that even when in the hand they may be satisfactorily identified, and it is quite impossible to describe them so that from color alone they may be recognized in the field. Fortunately, the calls of our commoner species are so unlike that, when learned, there will be no difficulty in naming their authors.

To say that the Least Flycatcher is five and a half inches long, olive-green above and grayish white below, does not aid one in distinguishing it from several of its cousins; but when I add that its call is a snappy *chebèc*, *chebèc*, the bird will be known the first time it is heard. It is this call which has given the bird its common name.

The Chebec comes to us in the spring, about April 25, and remains until September.

You will rarely find two members of the same family with more different dispositions than those of the King-bird and Wood Pewee. Their natures might symbolize war and peace, so combative is the Kingbird, so gentle the

WOOD PEWEE

Length, 6 50 inches. Upper parts dusky olive-green ; under parts whitish, washed with dusky ; lower mandible yellowish.

Pewee. As so often happens among birds, their voices
are in keeping with their temperaments. The soft,

Wood Pewee,   dreamy *pee-a-wee* or *pee-a-wee peer* of
Contopus virens.   the Pewee is as well suited to its char-
Plate XXXIII.   acter as the harsh, chattering cries of
victory are to the Kingbird's.

The Pewee is the last of our more common Flycatchers to come from the South, arriving about May 10,
and, like the Chebec, remaining until October. It is less
social than either the Chebec or the Phœbe. Forests
are its chosen haunts, but occasionally it is found on well-
shaded lawns and roadsides.

The Pewee's nest rivals the Hummingbird's in beauty.
It is a coarser structure, composed of fine grasses, rootlets,
and moss, but externally is thickly covered with lichens.
Usually it is saddled on a limb from twenty to forty feet
above the ground. The eggs, three or four in number,
are white, with a wreath of dark brown spots around the
larger end.

LARKS. (FAMILY ALAUDIDÆ.)

This family contains the true Larks, birds with long
hind toe nails, and a generally brown or sandy colored
plumage, the Skylark being a typical species. There are
some one hundred species of Larks, but of these only the
Horned Lark and its geographical varieties are found
in this country.

The variation in color shown by the Horned Lark
throughout its range is remarkable. From the Mexican

Horned Lark,   tableland northward to Labrador and
Otocoris alpestris.   Alaska no less than eleven different
Plate XXXIV.   geographical races are known, each one
reflecting the influence of the conditions under which it
lives, and all intergrading one with another. Only two of

HORNED LARK.

Length, 7·75 inches. Upper parts brownish and sandy ; front and sides of crown, sides of throat, and breast-patch black ; forehead, line over eye, and throat pale yellow ; breast dusky, belly white, tail black, outer feathers margined with white.

these races are found in the eastern United States, the Horned Lark and the Prairie Horned Lark. The former visits us in the winter; the latter occurs at all seasons, but during the summer is found only in certain regions. At this season it inhabits the upper Mississippi Valley, whence it extends eastward through northwestern Pennsylvania and central New York to western Massachusetts. From October to April it may be found with the Horned Lark as far south as South Carolina. The two birds differ in size and color. The Horned Lark's wing averages 4·27 inches in length, the Prairie Lark's wing averages but 4·08 inches in length; the former's forehead and eye-line are yellow, the latter's white.

Horned Larks are eminently terrestrial, rarely if ever choosing a higher perch than a fence. When on the ground they do not hop, but walk or run. When flushed they take wing with a sharp, whistled note, but often return to the place from which they started. When nesting, they may be found in fields, pastures, and plains in scattered pairs, but during the winter they are associated in flocks, which resort to the vicinity of the seacoast or large open tracts in the interior. The nest is, of course, built on the ground. The eggs, three or four in number, are pale bluish or greenish white, minutely and evenly speckled with grayish brown.

The Horned Lark, like its famous relative and many other terrestrial species, sings while on the wing, soaring high above the earth, and often repeating its song many times before alighting. The effort is worthy of better results, for the bird's song is simple and unmusical.

CROWS, JAYS, ETC. (FAMILY CORVIDÆ.)

There are systematists who think that the members of this family should hold the place usually assigned the Thrushes, at the head of the class *Aves*. Leaving out of the case anatomical details whose value is disputed, we might object to a family of songless birds being given first rank in a group whose leading character is power of song. But while Crows and Jays may, from a musical standpoint, be considered songless, no one can deny their great vocal powers. Song, after all, does not imply high rank in bird-life, and some of the sweetest singers (among others, some Snipe, and the Tinamous and Wood Quail of South America) are not members of the suborder of Song Birds.

If, however, the relative intelligence of the two families be taken into account, there can be no doubt that *Corvidæ* fully deserve to be considered the most highly developed of birds. How many tales are told of the human actions of the Raven, Rook, Jackdaw, Magpie, Jay, and Crow!

Of the two hundred members of this family, six inhabit eastern North America, by far the most common being the Crow. No one of our birds

American Crow,
Corvus americanus.

is better known, and still, how ignorant we are of his ways! I am not sure that he does not know more about ours. We have not even recorded his notes, for, in spite of the current opinion that the Crow's calls are restricted to *caw*, he has an extended vocabulary. I am not aware that he ever ascends to the height of a love song, but that he can converse fluently no one who has listened to him will question. Of the variants of *caw*, each with its own significance, there seems no end; but if you would be impressed with the Crow's eloquence you must hear him when, in the fancied privacy of his own flock or family, he discusses the affairs of the day. His notes then are low, and so varied in tone that one can not doubt their conversational character.

During the winter Crows assemble in large flocks containing many thousand individuals, who nightly return to some roost, which perhaps has been frequented for years. In March they begin to pair and the nest is constructed early in April. It is a bulky affair of sticks, lined chiefly with grapevine bark, and is placed in a tree, usually about thirty feet from the ground. The four to six eggs are bluish green, thickly marked with shades of brown.

Crows share with Hawks the reputation of being harmful birds. That they do much damage in the cornfield is undeniable, but, after the examination of nine hundred Crows' stomachs, Dr. Merriam, of the Department of Agriculture, states that the amount of good done by the Crow in destroying grasshoppers, May beetles, cutworms, and other injurious insects, exceeds the loss caused by the destruction of corn. Moreover, if the corn be tarred before planting, the Crows will not touch either the kernel or young sprout. The corn should first be soaked in water overnight, and then placed in a vessel containing enough soft tar to coat each kernel. It should then be rolled in plaster of Paris or wood ashes, so that it can be more easily handled.*

The Blue Jay, in his uniform of blue and white, is so brightly colored, so large (he is nearly twelve inches in length), and often so noisy, that every one knows him.

---

* See Barrows and Schwarz, The Common Crow, Bulletin No. 6, United States Department of Agriculture, Division of Ornithology and Mammalogy.

Like the Crow, he is with us throughout the year. During the summer he is not very common, and is remarkably quiet, but in September and October migrants arrive from the North, and the birds are then abundant in bands.

Blue Jay,
Cyanocitta cristata.

These bands roam about the country like a lot of schoolboys out chestnutting, pausing wherever they find acorns and chestnuts abundant, or leaving their feast to worry some poor Owl whose hiding place they have discovered.

The Blue Jay's best friend could not conscientiously call him a songster, but as a conversationalist he rivals the Crow. I have yet to discover a limit to his vocabulary, and, although on principle one may ascribe almost any strange call to the Blue Jay, it is well to withhold judgment until his loud, harsh *jay! jay!* betrays the caller's identity. Not content with a language of his own, he borrows from other birds, mimicking their calls so closely that the birds themselves are deceived. The Red-shouldered, Red-tail, and Sparrow Hawks are the species whose notes he imitates most often.

The Blue Jay nests in the latter part of May, building a compact nest of rootlets in a tree ten to twenty feet from the ground. The eggs are pale olive-green or brownish ashy, rather thickly marked with varying shades of cinnamon-brown.

ORIOLES, BLACKBIRDS, ETC. (FAMILY ICTERIDÆ.)

The popular names of many of our birds were given them by the early colonists because of their fancied resemblance to some Old World species. The fact that some of these names are incorrect and misleading has been pointed out scores of times, but they are now as firmly fixed as the signs of the zodiac.

Thus the Robin is not a Robin but a true Thrush, the Meadowlark is not a Lark but a Starling, and the Orioles are not Orioles at all, but members of a distinctively American family having no representatives in the Old World. This family contains one hundred and fifty species, of which nearly one third belong in the genus *Icterus*. The prevailing colors of the birds of this genus are orange and black, hence their resemblance to the true Orioles (genus *Oriolus*) of the Old World.

Our Baltimore Oriole is a worthy representative of a group remarkable for its bright colors. It is to these

Baltimore Oriole, same colors that the bird owes not only *Icterus galbula.* its generic but its specific designation, Plate XXXV. orange and black being the livery of Lord Baltimore, after whom the bird was named.

The Baltimore Oriole, or, as it is also called, Firebird, Golden Robin, or Hangnest, winters in Central America, and in the spring reaches the latitude of New York city about May 1. I always look for it when the cherry trees burst into blossom, and at no other time does its beautiful plumage appear to better advantage than when seen against a background of white flowers. To the charm of beauty it adds the attraction of song, a rich, ringing whistle, which can be more or less successfully imitated, when the bird immediately responds, challenging the supposed trespasser on his domain.

The Baltimore's nest is a bag about five inches deep and three inches in diameter, woven of plant-fibers, thread, etc., and suspended from the terminal portion of a limb, generally of an elm tree. The four to six eggs are white, singularly scrawled with fine black lines, and with a few spots or blotches.

The Orchard Oriole is neither so common nor so gayly dressed as his brilliantly colored relative, and, being fonder of orchards than lawns and elm-shaded highways,

BALTIMORE ORIOLE.

Length, 7 50 inches. *Male*, crown, upper back, and throat black ; lower back, outer tail-feathers, breast, and belly rich orange. *Female*, upper parts mixed black and yellowish, rump and tail dirty yellow ; under parts dusky yellow.

is not so well known. The female is especially easy to overlook, her suit of plain olive-green closely harmoniz-

Orchard Oriole,
Icterus spurius.
Plate XXXVI.

ing with the leaves in color. Young males at first exactly resemble her, but the following spring return, wearing their father's black cravat. In this plumage they might readily be taken for another species, so little do they resemble their parents in appearance. The adult chestnut and black plumage is not fully acquired until the second, or perhaps even the third spring.

The Orchard Oriole winters in Central America, and in the summer is found throughout the eastern United States from the Gulf of Mexico to Massachusetts. It arrives from the South about May 1, and is one of the first birds to leave in the fall, rarely being seen after September 1. Nesting is begun late in May. The nest is pensile, but not so deep as that of the Baltimore Oriole, having more the proportions of a Vireo's nest. It is composed entirely of freshly dried greenish grasses, and is suspended from near the extremity of a branch at a height of fifteen to twenty feet. The three to five eggs are bluish white, spotted, blotched, and scrawled with black.

The song of the Orchard Oriole resembles that of his orange-and-black cousin, but is far richer in tone and more finished in character.

The male Redwing, with his black uniform and scarlet epaulets, is a familiar inhabitant of our marshes, but

Red-winged Blackbird,
Agelaius phœniceus.

many who know him are not acquainted with his very differently attired mate. She wears a costume which above is black streaked with buff and rust-color, and below is striped dingy black and white, and is much more retiring than her conspicuous husband. Her place is low in the bushes or among the reeds near the nest with its pale

PLATE XXXVI.

ORCHARD ORIOLE.

Length, 7 30 inches. *Adult male*, crown, back, and throat black, rest of body chestnut. *Young male*, upper parts olive-green ; throat black, rest of under parts yellowish *Female*, similar, but black on throat replaced by yellowish

167

blue eggs, so singularly scrawled with black. He perches on the topmost branch of a neighboring tree, and doubtless supposes he is guarding his home below, when in truth he is advertising his treasure to every passer-by.

The Redwing's liquid *kong-quĕr-rēē* is pleasantly suggestive of marshy places, but it is his early spring music for which we should chiefly value him. The first Robins or Bluebirds are somewhat unreliable signs of spring. They are such hardy birds that it requires very little encouragement from a February sun to send a few skirmishers northward. We can not be sure whether they represent the advance guard or are individuals who have had the courage to winter with us. But when early in March the Redwings come, then we know that the tide of the year has turned. With perennial faith in the season they come in flocks of hundreds, singing their springtime chorus with a spirit that March winds can not subdue.

About the time the Redwings come, late in February or early in March, we may expect the Purple Grackles

Purple Grackle,
Quiscalus quiscula.
Plate XXXVII.

or Crow Blackbirds. They migrate in large flocks, and their chorus singing is quite as inspiring as the springtime concerts of the Redwing. There are two kinds of Crow Blackbirds, known as the Purple Grackle and the Bronzed Grackle. The former has iridescent bars on the back and in the Northern States is found only east of the Alleghanies and south of Massachusetts; the latter has the back shining, brassy, bronze, without iridescence, and in the nesting season inhabits the country west of the Alleghanies and north of Connecticut. The females of both species are smaller and duller than the males.

Grackles are among the few of our land birds who live in flocks all the year. They pass the winter and migrate in larger companies, but when resting are in smaller

PURPLE GRACKLE.

Length, male, 12·50 inches ; female, 11·00 inches. *Male*, head, neck, throat, and breast bright metallic blue, purple, or green ; back with iridescent bars ; belly paler ; eye pale yellow. *Female*, much duller than male.

bands or colonies. They generally select a pine grove, often choosing one in a cemetery, park, or other locality where they will not be disturbed. This may result in a scarcity of food when the young are born, but, rather than abandon a locality which experience has proved to be safe, they make long journeys in search of food for their nestlings. By watching the old birds one may then easily learn where they live. Their flight is direct and somewhat labored, and when going only a short distance they "keel" their tail-feathers, folding them upward from the middle, an action which renders Grackles conspicuous and easily identifiable when on the wing. On the ground they strut about with a peculiar walk, which, in connection with their yellowish white eye, adds to the singularity of their appearance.

The Grackle's nest is a bulky, compact structure of mud and grasses. It is usually placed in trees, twenty to thirty feet from the ground, but the bird may sometimes nest in bushes or even in a Woodpecker's deserted hole. The three to six eggs are generally pale bluish green, strikingly spotted, blotched, or scrawled with brown and black. But one brood is raised, and when the young leave the nest they roam about the country in small bands, which later join together, forming the enormous flocks of these birds we see in the fall.

The Bobolink's extended journeys and quite different costumes have given him many *aliases*. Throughout his breeding range, from New Jersey to Nova Scotia, and westward to Utah, he is known while nesting as the Bobolink. In July and August he loses his black, buff, and white wedding dress, and gains a new suit of feathers resembling in color those worn by his mate, though somewhat yellower. This is the Reedbird dress, and in it he journeys nearly four

Bobolink,
Dolichonyx
oryzivorus.
Plate XXXVIII.

In June Meadows

PLATE XXXVIII.

BOBOLINK.

Length, 7 25 inches. *Male, in summer,* nape buff ; shoulders and rump whitish ; crown and under parts black *Female, young, and male in winter,* sparrowlike ; upper parts black, brownish, and buffy ; under parts yellowish white.

thousand miles to his winter quarters south of the Amazon.

The start is made in July, when he joins flocks of his kind in the northern wild-rice (*Zizania aquatica*) marshes. Late in August he visits the cultivated rice fields of South Carolina and Georgia, and it is at this season we so often hear the metallic *tink* of passing migrants. The rice is now in the milk, and the Ricebirds, or Ortolans, as they are called in the South, are so destructive to the crop that it is estimated they directly or indirectly cause an annual loss of $3,000,000. Some birds linger as far north as New York until October 1, but by this time the leaders of the south-bound host have reached Cuba, where they are called *Chambergo*. From Cuba they pass to the coast of Yucatan, and thence southward through Central America or to the island of Jamaica, where, because of their extreme fatness, they are known as Butterbirds. From Jamaica they go to the mainland, either of Central America, or by one continuous flight of four hundred miles to northern South America, thence traveling southward to their winter home.

The northward journey is begun in March or April, and about the 25th of the latter month the vanguard reaches Florida. It is composed only of males, now called Maybirds, all in full song. Let any one who knows the Bobolink's song imagine, if he can, the effect produced by three hundred birds singing together!

About May 1 Bobolinks reach the vicinity of New York city. The females soon follow the males, and early in June the birds are nesting. This is the glad season of the Bobolink's year. For ten months he has been an exile, but at last he is at home again, and he gives voice to his joy in the jolliest tinkling, rippling, rollicking song that ever issued from bird's throat.

In the fields made merry by the music of Bobolinks one

PLATE XXXIX.

MEADOWLARK.

Length, 10·75 inches. Upper parts black, brown, and buff ; under parts yellow, a black crescent on the breast, sides streaked with black ; outer tailfeathers white.

is almost sure to find Meadowlarks. They are strong-
legged walkers, and spend all their time while feeding
Meadowlark,   on the ground.   Like all terrestrial,
Sturnella magna.   protectively colored birds, they often
Plate XXXIX.   try to escape observation by hiding in
the grasses rather than by flying. When perched in a
tree or other exposed position, they are among the shyest
of our smaller birds, rarely permitting a near approach ;
but when they fancy themselves concealed on the ground
they sometimes " lie as close " as Bob-whites. When
flushed they fly rapidly, alternately flapping and sailing,
showing as they fly the white feathers on either side of
their tail. These feathers are the Meadowlark's best field
character. They are very conspicuous when he is on the
wing, and, when perching, if he is alarmed or excited, he
exposes them by nervously flitting or twitching his tail.
This movement is generally accompanied by a single
nasal call-note, which changes to a rolling twitter as the
bird takes wing. Neither of these notes give any indication of the sweetness of the bird's song, a high musical
whistle, clear as the note of a fife, sweet as the tone of a
flute. It is subject to much variation both individual
and local, but the song I oftenest hear in northern New
Jersey may be written :

When singing, the birds usually perch in an exposed position, generally choosing the topmost branches of a tree
or a dead limb.

The Meadowlark's nest is placed upon the ground, as
a rule, in a tuft of grasses which is arranged to form a
dome over it. The eggs, four to six in number, are laid
about May 15, and in color are white, spotted or speckled
with cinnamon or reddish brown.

COWBIRD.

Length, 7·90 inches. *Male*, head and neck all around dark coffee-brown ; rest of plumage glossy greenish black. *Female*, dirty brownish gray ; throat whitish.

175

Occasionally Cowbirds are seen during the winter near New York city; but, as a rule, they retire farther

Cowbird,
Molothrus ater.
Plate XL.

south at this season, and are first observed there in the spring about March 20. They do not come in large flocks, but singly or in small bands. The male may now be seen perched in an exposed position on a treetop, calling his long-drawn-out, glassy *kluck, tsē-ē-ē.* Later, when wooing the female, he utters a curious, gurgling note, resembling the sound made by pouring water rapidly from a bottle, and accompanying it by motions which suggest extreme nausea. We often see these birds feeding near cattle in the pastures, always in small flocks, for they do not pair nor even construct a nest, the female laying her egg in the nest of another and generally smaller species. Few birds seem aware of the imposture, and not only do they incubate the egg but they may attend to the demands of the young Cowbird at the expense of their own offspring, who sometimes die of starvation. Even after leaving the nest the young parasite continues its call for food, and when seeing a Maryland Yellowthroat, or some other small bird feeding a clumsy fledgling twice its size, one wonders it does not detect the deception. The better we know birds the more strongly are we impressed with their individuality. To one who has no friends in feathers it seems pure fancy to endow some insignificant "Chippy" with human attributes; but in reality there are as clearly defined characters among birds as among men. To be convinced of the truth of this statement we have only to compare the Cowbird, a thoroughly contemptible creature, lacking in every moral and maternal instinct, with the bird who constructs a well-made nest, faithfully broods her eggs, and cares for her young with a devotion of which mother love alone is capable.

PLATE XLI.

SONG SPARROW.

Length, 6.25 inches. Upper parts chestnut, gray, and black ; under parts
white, streaked with chestnut and black ; outer tail-feathers shortest

177

13

SPARROWS, FINCHES, ETC. (FAMILY FRINGILLIDÆ.)

This, the largest family of birds, contains between five hundred and fifty and six hundred species, and is represented in all parts of the world except the Australian region. Sparrows are the evergreens among birds. When the leaves have fallen from the chestnut, oak, and maple, the hemlock, pine, and cedar are doubly dear. So, when the Flycatchers, Warblers, and Thrushes have left us, the hardy Sparrows are more than usually welcome. Feeding largely on seeds, which their strong, stout bills are especially fitted to crush, they are not affected by the changes in temperature which govern the movements of strictly insectivorous birds.

Some species are with us throughout the year, some come from the South in early spring and remain until snow falls, others come from the far North to pass the winter; so that at no season of the year are we without numbers of these cheery birds. Fortunately, some of our best songsters are members of this family. Their music is less emotional than that of the Thrushes, but it has a happier ring—the music for every day.

It is the Song Sparrow who in February opens the

Song Sparrow,
Melospiza fasciata.
Plate XLI.

season of song, and it is the Song Sparrow who in November sings its closing notes ; nor, except during a part of August, has his voice once been missing from the choir.

His modest chant always suggests good cheer and contentment, but heard in silent February it seems the divinest bird lay to which mortal ever listened. The magic of his voice bridges the cold months of early spring ; as we listen to him the brown fields seem green, flowers bloom, and the bare branches become clad with softly rustling leaves.

Neighbors

PLATE XLII.

SWAMP SPARROW.

Length, 5·90 inches. *Summer plumage,* crown bright chestnut ; back black, brown, and buff ; breast grayish ; belly white ; sides brownish. *Winter plumage,* similar, but crown streaked with chestnut-brown, black, and gray.

You can not go far afield without meeting this singer. He is not only our commonest Sparrow, but one of our commonest birds. Generally you will find him on or near the ground at the border of some undergrowth, and if there be water near by, preferably a meadow brook, his presence is assured. When flushed he will doubtless make for the nearest thicket, "pumping" his tail, as Thompson expressively says, in describing his somewhat jerky flight. Now he questions you with a mildly impatient *chimp* or *trink*, a call-note not to be mistaken for that of any other species, when once you have learned it. Equally diagnostic is the bird's spotted breast with one larger spot in its center.

The Song Sparrow's nest is usually placed on the ground, but sometimes a bush may be chosen for a nesting site. The eggs, four or five in number, are bluish white, thickly marked with reddish brown. The Song Sparrow rears three broods each year, the nesting season lasting from May to August.

The Swamp Sparrow, a well-named cousin of the Song Sparrow, resembles his relative in his fondness for

Swamp Sparrow,
Melospiza georgiana.
Plate XLII.

the vicinity of water and habit of taking refuge in low cover. He is a true marsh or swamp bird, and is particularly abundant in large marshes. His call is an insignificant *cheep*, while his song is a simple, sweet, but rather monotonous *tweet-tweet-tweet*, repeated many times and occasionally running into a trill.

The Swamp Sparrow nests from northern Illinois and Pennsylvania northward to Labrador. Its nest and eggs resemble those of the Song Sparrow. It is migratory in the northern part of the range, and is rare in winter north of southern New Jersey.

Both the Song and Swamp Sparrow are, as we have seen, birds of the lowlands, though the latter also inhab-

PLATE XLIII.

FIELD SPARROW.

Length, 5·70 inches. Upper parts bright reddish brown and black ; under parts grayish white ; bill reddish brown.

its higher ground, but the two Sparrows now to be mentioned are birds of the uplands, rarely if ever living in low, wet places.

An old hillside pasture, dotted with young cedars or clumps of bushes, in which he may place his nest, is the

Field Sparrow,
Spizella pusilla.
Plate XLIII.

favorite home of the Field Sparrow. Here you may look for him early in April. He is a rather shy bird, who will fly some distance when alarmed, and then alight on a bare twig near or at the top of some bush or sapling. Very different this from the Song Sparrow's way of diving into a bush.

From his exposed position he watches you and gives you an equally good chance to watch him. Note the whitish, unstreaked breast, the reddish brown or sorrel crown, the gray face and whitish ring about the eye, and especially the pale brownish or flesh-colored bill. These are all good marks, and if now you can hear him sing his identity will be settled without question. His song is one of the most pleasing I know. It is very simple but very expressive, a sweet, plaintive *cher-wee, cher-wee, cher-wee, cheeo dee-e-e-e-e,* which goes straight to one's heart. It is sung most freely after sunset, and is in keeping with the peacefulness of the evening hour. At this time, too, the bird seems inspired to more than usual effort, and its ordinary song is often so elaborated and prolonged as to be scarcely recognizable.

The song season ends in the latter part of August, and, although the birds are with us until November, I have rarely heard them sing in the fall.

The Vesper Sparrow, Grass Finch, or Bay-winged Bunting—for he bears all three names—prefers more open grounds than the Field Sparrow selects. There is something free and spirited about this bird and its song which demands space for its proper development. No

PLATE XLIV.

VESPER SPARROW.

Length, 6 10 inches. Upper parts grayish, black, and brown ; breast and sides streaked with black and brown ; belly white ; lesser wing-coverts chestnut ; outer tail-feathers more or less white.

swamp or thicket will do for him, but in great broad fields
he is at home. If a roadway leads through his haunts,
Vesper Sparrow,   you may often see him on the ground
Pooecetes gramineus. ahead of you, and when he flies the
Plate XLIV.   white feathers shown on either side of
his tail will give you an excellent clew to his identity.
Probably he will fly on ahead a little way and alight
again in the road, or a longer flight may lead him to a
neighboring fence or the upper branches of a more distant tree. It is from positions of this kind that he most
often sings. With him song is evidently a matter of importance. He can not, like many birds, sing between the
mouthfuls of a meal, but ascending to his perch he gives
perhaps half an hour entirely to music, resting motionless
between the intervals of each song.

It is impossible to satisfactorily describe this song.
It resembles that of the Song Sparrow, but is finer and
wilder. It opens with one low note, followed by two
higher ones, while the Song Sparrow begins with three
notes, all of the same kind.

The Vesper Sparrow is migratory, coming to us with
the Field Sparrow early in April and remaining until
November. Its nest is placed on the ground, and the
bluish or pinkish white speckled eggs are laid early in
May.

It is strange, is it not, that the only bird we all detest
should also be the only one who insists on sharing our
homes with us. The House or English
House Sparrow, Sparrow, is a product of the times ; a
Passer domesticus. remarkably keen-witted bird, who, like
a noxious weed. thrives and increases where a less hardy
species could not exist.

This harsh-voiced little *gamin* soon detects and avoids
anything like a systematic attempt to entrap him, and,
being productive past all belief, seems likely to completely

PLATE XLV.

CHIPPING SPARROW.

Length, 5·35 inches. *Summer plumage,* forehead black ; crown bright chestnut ; back black, brown, and gray ; under parts grayish white ; bill black.
Winter plumage, similar, but crown like back ; bill brownish.

overrun the land. He was introduced into this country
in 1851, and in 1870 was found only in the cities of the
Atlantic States. Now he has spread over the greater
part of the United States and Canada.

If he were restricted to the cities we should have only
his never-ceasing, maddening chatter and our soiled walls
to complain of; but he has invaded not only the towns
and villages and the neighboring houses, but visits also
our grain fields and fruit orchards, our woods and marshes.
No effective method for his extermination has been de-
vised, and I fear we must accept the Sparrow as a penalty
for the shortsightedness and ignorance which permitted
us to meddle with the laws of Nature.

If we except this ever-present nuisance the Chippy
is the most domestic of our Sparrows. He seems thor-

Chipping Sparrow, oughly at home about our doorsteps; a
Spizella socialis. contented, modest little bird who ap-
Plate XLV. parently tries hard to believe in the
goodness of human nature, even though he meets with
but little encouragement. One wonders why he has not
long ago given up the attempt to make friends with us,
so rarely do we show any appreciation of his advances.
The house cat is Chippy's chief enemy. Crouching and
crawling, waiting and watching, she misses no opportunity
to pounce on an unsuspecting bird. It is surprising that
any escape. But each spring, about April 10, the Chippy
comes back to us after a winter in the cotton, corn, and
broom-sedge fields of the South, and soon we hear his
unpretentious, monotonous *chippy-chippy-chippy*, many
times repeated, and occasionally running into a grasshop-
perlike trill.

About a month later we may find further evidence of
his too often misplaced trust in a neat, hair-lined nest
built in the vines on the veranda or a neighboring tree.
The eggs are unexpectedly pretty, a bright blue or bluish

PLATE XLVI.

PLATE XLVI.

WHITE-THROATED SPARROW.

Length, 6·75 inches. *Adult*, lores and bend of wing yellow ; crown black and white ; back chestnut-brown, black, and buff ; throat white ; breast and sides grayish ; belly white. *Young*, similar, but crown more like back ; yellow markings duller.

187

green, spotted, chiefly at the larger end, with cinnamon-
brown or blackish markings.

Up to this time the Chippy has given us a good opportunity to see his chestnut cap and black forehead, but
when the nesting season is over he will change these for
a cap to match his coat, and with others of his kind gather
in old, weedy fields, remaining there until cold weather
drives him southward.

About the time of the first frost a new Sparrow will
appear in the hedgerows and thickets and the undergrowth of the woods. The white patch
White-throated
Sparrow, on his throat may aid in his identifica-
Zonotrichia albicollis. tion as the White-throated Sparrow, a
Plate XLVI. Northern bird who in the summer
nests from northern New England northward, and in
winter is found from southern New England to the Gulf
of Mexico.

He is disposed to be rather quiet for several days
after his arrival, and, beyond a few low notes addressed
to his companions, has little to say; but if you whistle
to him even a poor imitation of his song, nearly every
bird in the company will hop up from the tangle of
branches and, perching on the outer twigs, look for the
friends who called. Perhaps some may essay a tremulous
response, but for a week or more they will make few at-
tempts to sing. Later, you will hear the sweet, plaintive
notes that give to this bird the name Peabody-bird.

The White-throat's call-notes are a low *tseep* and a
very characteristic sharp *chink*, which has been well
likened by Mr. Bicknell to the sound of a marble cutter's chisel. At this season the White-throats roost together in flocks of varying size, and if you chance to be
near their home at bedtime you will hear this *chink* note
given as a " quarriers' chorus." Finally, as the gloom
deepens, it will cease, and from the dark depths of the

FOX SPARROW.

Length, 7·25 inches. Upper parts, wings, and tail bright reddish brown ; back and head mixed with a browner color ; under parts white and bright reddish brown.

thicket will come only the cozy, contented twitterings of the birds wishing one another good night.

The interest with which one examines a flock of White-throated Sparrows is intensified by the probability of

White-crowned Sparrow, *Zonotrichia leucophrys.* finding their distinguished relative the White-crown. In the Mississippi Valley he is often common, but in the Atlantic States he is sufficiently rare to be a character of importance.

The White-crown differs from the White-throat in having no white on the throat, which, like the breast, is gray, and in having the space before the eye black instead of yellow or white. In the fall his crown is brown, with a paler line through its center.

Near New York city I look for the White-crown in September and October, and again about May 15. Thompson describes its song as "like the latter half of the White-throat's familiar refrain, repeated a number of times with a peculiar sad cadence and in a clear, soft whistle."

Some fine day about the middle of March you may hear a song so unlike any you have ever heard, that be-

Fox Sparrow, *Passerella iliaca.* Plate XLVII. fore the singer ceases you will know you are on the verge of a discovery. The song is loud, exceedingly sweet, and varied. Its richness of tone seems to accentuate the bleakness of the bird's surroundings. It is a song for summer, not for leafless spring; but heard at this season it seems all the more attractive, and with pleasurable excitement you hasten toward the second growth, near the border of which the bird is perched. His large size and bright reddish brown upper parts readily distinguish him from other Sparrows, and, in connection with his spotted breast, give him a general resemblance to a Hermit Thrush, for which bird he is sometimes mistaken; but a

Ernest Seton Thompson

JUNCO.

Length, 6·25 inches. *Male*, upper parts, throat, and breast slate-color ; belly and outer tail-feathers white. *Female*, similar, but plumage more or less washed with brownish.

glance at his short, stout bill at once shows his family relationships, and you should have no difficulty in identifying
him as the Fox Sparrow.

A month later he will leave us for his summer home
in the far North, but in October and November his
ringing notes may again be heard as he pauses a day or
two on his journey southward.

After the Fox Sparrows go, our bird-life is reduced to
its winter elements—that is, permanent residents and win-

Junco, ter visitants. Of the latter the Junco

Junco hyemalis. or Slate-colored Snowbird is the com-

Plate XLVIII. monest and most generally distributed.
Although we call this bird a winter visitant, he is with
us nearly eight months in the year, arriving late in September and remaining until early May.

The Junco is one of the birds whose acquaintance can
be easily made. His suit of slaty gray, with its low-cut
vest of white, is not worn by any other of our birds; and
while some species show white outer tail-feathers in flight,
the Junco's seem to be more than usually conspicuous.

Except when nesting, Juncos associate in loose flocks
of from ten to fifty. Generally you will find them feeding on the ground near evergreens, into which, when disturbed, they will fly with a twittering note. If they are
excited by your appearance you will hear a sharp, kissing
call; but if unalarmed they will utter a rapidly repeated
chew-chew-chew, expressive of the utmost contentment.
In March and April, before leaving for their summer
home in northern New England or the crests of the
Alleghanies and Catskills, the Juncos sing a simple trill
or low, twittering warble. Modest in manner and attire,
there is nothing of especial interest in the Junco's habits,
and only bird-lovers can understand what a difference his
presence makes in a winter landscape. It brings a sense
of companionship; it is a link between us and Nature.

Wren the North Wind Doth blow.

PLATE XLIX.

TREE SPARROW.

Length, 6.35 inches. Crown bright chestnut ; back black, reddish brown, and buffy ; under parts grayish ; sides washed with brownish ; a blackish spot in the center of the breast.

14

The bird's cheery twitter is as welcome as a ray of sunlight on a cloudy day.

With the Juncos we may often find a company of Tree Sparrows or Winter Chippies. They resemble our familiar Chipping Sparrow, but the blackish dot in the center of their breasts is a good distinguishing mark.

Tree Sparrow,
Spizella monticola.
Plate XLIX.

Then, too, the true Chippies all leave for the South in November, while the Winter Chippies come in October and remain until April.

Tree Sparrows are sociable birds, with apparently the best of dispositions. They are usually found in small companies, each member of which seems to have something to say. Watch them feeding on an old weed stalk left uncovered by the snow. It bends beneath the weight of half a dozen birds, but, far from attempting to rob one another, they keep up a conversational chatter bespeaking the utmost good fellowship. *Too-lā-it, too-lā-it,* each one calls, and I have only to remember this note to bring clearly to mind a bright winter morning with the fresh snow crystals sparkling in the sunshine, and in the distance a tinkling chorus of Tree Sparrows at breakfast.

Another winter associate of the Junco's, and an intimate friend of the Tree Sparrow's, is the Redpoll, Redpoll Linnet, or, as he is sometimes called, Red-capped Chippy. The Redpoll nests in the far North, and the extent of his southern journeys depends very much upon the supply of food he finds in his winter wanderings. When there are seeds in abundance north of the United States, we do not see many of these birds, but if the larder fails they may come into New England in great numbers, and a few may venture as far south as Virginia. One can not tell, therefore, when to expect them, but it is well to be on the lookout from November to March.

Redpoll,
Acanthis linaria.
Plate L.

PLATE L.

REDPOLL.

Length, 5 30 inches. *Adult male*, crown bright red ; back brownish black and grayish ; throat black ; under parts white, streaked with black ; breast pink. *Adult female and young*, similar, but no pink on breast.

SNOWFLAKE.

Length, 6·90 inches. Upper parts brown and black ; wings and tail black and white ; under parts white ; breast and sides brownish.

With the Tree Sparrows and Juncos, Redpolls feed on
the seeds of plants left uncovered by the snow, and they
also include birch buds in their fare.

None of our winter birds better illustrate the flock-
ing habit than the Snowflakes, Snow Buntings, or, as they
Snowflake,
Plectrophenax nivalis.
Plate L. are also called, White Snowbirds. With
a uniformity of movement which would
put to shame the evolutions of the best-
drilled troops, they whirl over the snow-clad fields, wheel-
ing to right or left, as though governed by a single
impulse. Suddenly they swing downward into a weedy
field, alighting on the snow or ground, where they *run*—
not hop about—like little beach birds. Sometimes, it is
said, they sing on the wing while with us, but their usual
note is a low chirp. They are terrestrial birds, and, although they may often perch on fences or buildings, are
rarely seen in trees.

Snowflakes nest within the Arctic Circle, and, like
other of our winter birds that come from the far North,
are irregular in their movements. As a rule they do not
wander much south of Long Island and northern Illinois,
but occasionally they go as far as Virginia and Kansas,
and are thus among the possibilities which add so much
to the pleasure of winter days in the field.

The Crossbill is a possibility at any season. None of
our birds is more erratic in its migrations. As a rule, it
American Crossbill,
Loxia curvirostra
minor.
Plate LI. is found in the Middle States only between November and March, but I
have seen it in Central Park, New
York city, as late as May. In the
higher parts of the Alleghanies and in northern New
England it is resident throughout the year. Crossbills
usually wander as far south each winter as Connecticut,
but beyond this are of irregular occurrence.

They feed almost entirely upon the seeds of pines, and

PLATE LI.

AMERICAN CROSSBILL.

Length, 6·20 inches. *Adult male*, dull red ; back brownish ; wings and tail blackish. *Adult female and young*, greenish ; back more or less mottled with brownish ; the under parts grayish.

PINE GROSBEAK.

Length, 9·10 inches. *Adult male*, rose-pink ; back brownish ; lower belly gray ; wings and tail brownish black. *Adult female and young*, gray ; crown, upper tail-coverts, and breast washed with deep yellow.

are not often seen far from coniferous trees. Their singular bill might, at first glance, be considered misshapen, but if you will watch a Crossbill push his crossed mandibles beneath the scale of a pine cone, and with a quick twist force it off and secure the seed at its base, you will readily admit that for the bird's purposes his bill could not be easily improved.

In hunting for Crossbills it is a good plan to look through the woods for falling scales of pine cones, and when you see a shower of them whirling softly downward it behooves you to learn the cause of their descent. The birds often follow them to the ground, to secure the seeds which have dropped there.

Crossbills fly in compact flocks, and often utter a sharp, clicking note while on the wing. Their song is sweet and varied but not loud.

Pine Grosbeaks are among our rarer winter visitants. They come as far south as Massachusetts in vary-

Pine Grosbeak, ing numbers, and occasionally reach
Pinicola enucleator. Connecticut, but south of this point
Plate LI. are of very infrequent occurrence. At irregular intervals Pine Grosbeaks become abundant during the winter in New England, when, because of their size, they attract general attention. They usually resort to coniferous trees, upon the seeds of which they feed, but they also eat berries and buds, and are said to be especially fond of the fruit of the staghorn sumach.

No one seeing the Goldfinch or Yellowbird in his summer costume of gold and black would imagine that so

Goldfinch, dainty a creature could brave the storms
Spinus tristis. of winter; but late in the season, when
Plate LII. his home life is ended, he changes the gay wedding dress for a plainer suit, and joins the ranks of winter birds.

I wish that every one knew the Goldfinch. His gen-

PLATE LII

AMERICAN GOLDFINCH.

Length, 5·10 inches. *Adult male in summer*, crown black ; rest of body yellow ; wings and tail black and white. *Adult female and males in winter*, upper parts grayish brown ; crown yellowish ; under parts soiled whitish ; throat yellow.

tle ways and sweet disposition are never-failing antidotes for discontent. One can not be long near a flock of these birds without being impressed by the refinement which seems to mark their every note and action. They show, too, a spirit of contentment from which we may draw more than a passing lesson. *Hear me, hear me, dearie*, they call as they feed among the weeds or on the birch buds, and, no matter how poor the fare, they seem thankful for it. The seeds of the dandelion, thistle, and sunflower are among their favorites; and if you would attract Goldfinches as well as some other birds, devote a corner of your garden to sunflowers.

The meal finished, the birds launch into the air, and to the tune of a cheery *per-chic-o-ree, per-chic-o-ree*, go swinging through space in long, bounding undulations.

In April the males regain their bright colors, but they are evidently believers in prolonged courtship, and, although the nuptial dress is acquired so early, housekeeping is apparently not thought of until June. Then a neat home of bark and fine grasses, thickly lined with plant down, is placed in a bush or tree, five to thirty feet from the ground, and in it are laid three to six pale, bluish-white eggs.

Now the song season has reached its height. Chorus singing has been abandoned. Each bird has become an inspired soloist, who, perched near his home or flying in broad circles about it, pours forth a flood of melody. It is an exceedingly attractive song, sweet and varied and suggesting a Canary's, but still is no more like it than a hothouse is like a tropical forest.

Creak, creak, the notes are clear but faint, and may come from any place beyond arm's reach.

Purple Finch,
Carpodacus purpureus.
Plate LIII.

They are the Purple Finch's flight-calls; one might think his wing-joints needed oiling. Alighting on the topmost twig of a

Ernest Seton Thompson

PLATE LIII.

PURPLE FINCH.

Length, 6·20 inches. *Adult male,* rose-pink ; back brownish ; lower belly white ; no white in wings. *Adult female and young,* upper parts streaked brownish and grayish ; under parts white, streaked with brownish ; bill rounded on top ; a tuft of bristly feathers over the nostrils.

forest tree, he utters a low, wild, questioning whistle. With crown-feathers slightly erect he seems alert and restless, and before we can fairly see him is off again to parts unknown.

Purple Finches, in small companies, may often be seen feeding near the ground with Goldfinches, but if alarmed they soon return to the tree tops. The old males may be known by their pinkish red color, which is bright-est on the head and breast, and fades to brownish on the lower back and tail and white on the belly. The young males and females are Sparrowlike in appearance, the upper parts being dark grayish brown, the under parts white, streaked with dusky. A whitish line passing over the eye is a characteristic mark.

During the winter Purple Finches are irregularly dis-tributed throughout most of the Eastern States, but in summer they are not found south of northern New Jer-sey. They now become more social and may nest in our gardens. Generally a coniferous tree is selected, and the nest of twigs, grasses, and rootlets is placed at a height of about twenty feet. The eggs, four to six in number, are blue, spotted with dusky about the larger end.

Count yourself fortunate if a Purple Finch makes his home near yours. He may appropriate a few buds and blossoms, but he will repay you with music and leave you his debtor. His song is a sweet, flowing warble; music as natural as the rippling of a mountain brook.

Some morning early in May you may meet the Rose-breasted Grosbeak, just returned from a winter's sojourn

Rose-breasted Grosbeak,

Zamelodia

ludoviciana.

Plate LIV.

in South America. Perhaps his fame will have preceded him, when you will in a measure be prepared for his charms of song and plumage, and so miss the keener pleasure of surprise; but to me he appeared as a revelation, and after fifteen years I still

PLATE LIV.

ROSE-BREASTED GROSBEAK.

Length, 8·10 inches. *Adult male*, crown and back black ; rump white ; throat black ; breast rose-red ; belly white. *Adult female*, upper parts dark brown and buff ; a white line over eye ; under parts buffy, streaked with brownish ; under wing-coverts orange.

find it difficult to believe that, unknown to me, this beautiful creature could long have been an inhabitant of my woods.

The Grosbeak prefers young second growths, with a liberal proportion of oaks. In one of these trees he will doubtless build his nest, a structure so lightly made that one can almost see the blue, spotted eggs from below. The male is not only an ardent lover but an admirable husband, and, unlike most brightly attired birds, shares with his mate the task of incubation, and, it is said, sings while on the nest. His mate is so unlike him in color that few would suspect their relationship. She suggests an overgrown female Purple Finch, with the eye-stripe especially prominent; but if you should chance to see the under surface of her wings, you would find that they were lined with gold. However, the call-notes of both sexes are alike—a sharp, characteristic *peek*, which you will have no difficulty in recognizing after you have learned it.

The Grosbeak's song will remind you of a Robin's, but it is in truth a much higher type of bird music. It is a joyous carol, expressive of a happy disposition and a clear conscience.

The Towhee, or Chewink, is an important member of any bird community. He comes early—April 20 may

Towhee,

Pipilo

erythrophthalmus.

Plate LV.

find him with us—and he stays late, sometimes remaining until November 1. During this period there is not an hour of the day when you can not find a Chewink if you know how to look for him. At midday you will perhaps have to summon him by a whistled *to-whée* from the depths of his bushy home on the border of a wood or thicket; but he will soon respond, and with a *fluff-fluff* of his short, rounded wings, fly jerkily up to inquire what's wanted.

PLATE LV.

TOWHEE.

Length, 8·35 inches. *Adult male*, upper parts, throat, and breast black ; belly white ; sides reddish brown. *Adult female*, similar, but black replaced by brownish.

Some birds, such as the Red-eyed Vireo, can sing just as well while hunting food as at any other time ; in fact, I do not remember ever seeing a Red-eye pause long in its search for insects—song and search go on together. But with the Chewink singing is a serious matter, not to be associated with the material question of food ; so, when singing, he abandons the dead leaves he has been tossing about so vigorously, and, mounting a perch, becomes an inspired if not gifted musician. *Sweet bird, sing*, a friend writes it, the " sing " being higher, sustained, and vibrant. To this there is often a refrain which suggests an answering, tremulous *I'll try.*

Matins or vespers over, the Chewink returns to the ground and resumes his occupation of scratching among the leaves for breakfast or supper, as the case may be.

The Chewink's nest is placed on the ground, often in dried grass, beneath a tangle of running wild blackberry. The eggs, four or five in number, are white, finely and evenly speckled with reddish brown.

There are three birds who sing not only through the heat of midsummer but are undaunted by the warmth of a midday sun. They are the Wood Pewee, the Red-eyed Vireo, and the Indigo-bird or Bunting. The Pewee and Vireo, singing dreamily from the shady depths of a tree, carry the air to the hummed accompaniment of insects; but the Bunting, mounting to an upper branch, gives voice to a tinkling warble, more in keeping with the freshness of early morning than the languor of noon. *July, July, summer-summer's here ; morning, noontide, evening, list to me,* he sings so rapidly that human tongue can scarce enumerate the words fast enough to keep pace with him. The Indigo-bird is in song when he comes to us from the South early in May, but it is not until other

Indigo Bunting,
Passerina cyanea.

singers have dropped from the chorus that his voice becomes conspicuous.

Not far away his mate is doubtless sitting on her bluish white eggs in a nest low down in the crotch of a bush. He in his deep indigo costume may be easily identified, but she is a dull brownish bird, about the size of a Canary, sparrowlike in appearance, though with unstreaked plumage, and a difficult bird to name, even when you have a specimen in your hand, while in the bush, if silent, she is a puzzle. But she is far too good a mother not to protest if you venture too near her home, and her sharp *pit* or *peet* usually calls her mate, whom you will recognize at once.

The Cardinal is about the size of a Towhee, with plumage which, except for a black throat, is almost

Cardinal,
Cardinalis cardinalis.

wholly rosy red. Seeing a mounted Cardinal, one might imagine that he was a conspicuous bird in life and easy to observe; but the truth is that, in spite of his bright colors, the Cardinal is a surprisingly difficult bird to see. You may often hear his sharp, insignificant *tsip* without catching a glimpse of the caller, so well can he conceal himself. His olive-brown mate is, of course, even more difficult to find, and when you do see her you would hardly suspect the relationship were it not for her actions and the striking crest worn by both sexes.

The Cardinal's song is a rich, sympathetic whistle. His mate also sings at times, and I carry in my memory a musical courting I once observed, in which a pair of these beautiful birds were the actors. The song begins with *whee-you, whee-you*, long-drawn notes, which are followed by a more rapid *hurry, hurry, hurry; quick, quick, quick*, and other notes difficult of description. The Cardinal is a bird of the Southern rather than of the Northern States, and is rarely seen north of New

York city. It is, however, a permanent resident throughout its range, and to one who associates it with magnolias
and yellow jessamine it seems strangely out of place amid
snowy surroundings.

The Cardinal builds its nest about four feet from the
ground in thickets, laying three or four eggs, which are
white or bluish white, speckled and spotted with grayish
or reddish brown.

In the Mississippi Valley and westward there are several members of this family who are rarely found east

Lark Finch, of the Alleghanies. Prominent among

Chondestes them is the Lark Finch, a handsome

grammacus. bird, about six and a quarter inches
long, with ear-coverts and sides of the crown chestnut,
the back grayish brown streaked with black, the outer
tail-feathers tipped with white, and the under parts
white, with a single black spot in the center of the
breast.

This is a migratory bird, arriving in southern Illinois
about the middle of April and remaining until September
or October. Mr. Ridgway, in his Birds of Illinois, says
that its favorite resorts are "fertile prairies and meadows
adjoining strips or groves of timber. In Illinois it evinces
a special fondness for cornfields, in which it builds its
nest at the foot of the stalks, while the male sings from
the fence or the top of a small tree by the roadside."

Its song, the same writer continues, is "composed of
a series of chants, each syllable rich, loud, and clear, interspersed with emotional trills. At the beginning the
song reminds one somewhat of that of the Indigo-bird
(*Passerina cyanea*), but the notes are louder and more
metallic, and their delivery more vigorous. Though
seemingly hurried, it is one continuous gush of sprightly
music; now gay, now melodious, and then tender beyond
description—the very expression of emotion."

PLATE LVI.

DICKCISSEL.

Length, 6·00 inches. *Adult male*, back black, chestnut, and grayish ; lesser
wing-coverts bright chestnut ; chin white ; throat black ; breast yellow ;
belly white. *Adult female*, upper parts streaked black and grayish ; throat
white ; breast yellowish, with black streaks ; belly white.

15

Some thirty or forty years ago the Dickcissel, or Black-throated Bunting, was a locally common bird in the Middle Atlantic States. Now it is rarely found east of the Alleghanies, and even in the Mississippi Valley its range is becoming restricted, and it is of irregular distribution.

Dickcissel,
Spiza americana
Plate LVI.

It migrates in large flocks, the males in the spring being several days in advance of the females. About May 1 it reaches the latitude of Chicago, and by the middle of the month is mated. The nest is placed on the ground, or in low trees or bushes; the eggs, four or five in number, are pale blue.

In the work previously quoted from, Mr. Ridgway writes of this species: "While some other birds are equally numerous, there are few that announce their presence as persistently as this species. All day long, in spring and summer, the males, sometimes to the number of a dozen or more for each meadow of considerable extent, perch upon the summits of tall weed stalks or fence-stakes, at short intervals, crying out: *See, see— Dick, Dick Cissel, Cissel;* therefore 'Dick Cissel' is well known to every farmer's boy as well as to all who visit the country during the season of clover blossoms and wild roses, when 'Dame Nature' is in her most joyous mood."

TANAGERS. (FAMILY TANAGRIDÆ.)

The Tanagers, numbering some three hundred and fifty species, are found only in America. Their home is in the tropics, where they are among the most abundant of birds. But two species reach the eastern United States, the Summer Redbird of the South and our Scarlet Tanager, both worthy representatives of a group of birds which in brilliancy of color rival even the Hummingbirds. The male Scarlet Tanager, with fire-red body and jet-black wings and tail, is the most brightly plum-aged of our birds. Seen against a

Scarlet Tanager,
Piranga erythromelas.

leafy background, light seems to radi-ate from his glowing feathers. But the female, clad in dull olive-green, is so in harmony with the color of her surroundings that she is not easily discovered. The young male at first resembles his mother, but has blackish wings and tail, and does not acquire the full scarlet and black plumage until the following spring. After the nesting season is over the male exchanges the nuptial dress, which has rendered him so conspicuous, for a costume similar to that worn by the young male.

The Scarlet Tanager spends the winter in Central and South America with his numerous relatives, and in the spring reaches the latitude of New York city about May 5, remaining until October. It frequents both high and low woods, but prefers rather open growths of white oak. Its nest is usually placed on the horizontal branch of an oak limb. The three or four eggs are pale greenish blue, with numerous reddish brown markings.

The Tanager's call-note is a characteristic *chip-churr* ; his song is not unlike the Robin's, but is not so free and ringing. Mounting to the topmost branch, often of a dead or partially dead tree, he sings, *Look-up, way-up, look-at-me, tree-top,* and with frequent pauses repeats the invitation.

SWALLOWS. (FAMILY HIRUNDINIDÆ.)

Primarily, Swallows are remarkable for their power of flight. Their long, bladelike wings show how well they are fitted for life in the air ; their small feet, on the other hand, are of little service except in perching, and give evidence of the effect of disuse (see Fig. 6).

The aërial ability of Swallows accounts for their wide distribution, the eighty known species being represented in all parts of the world. Only six of them inhabit the northeastern States, but they are so active and so easily observed, that they rank among our most abundant and best-known birds.

Swallows are eminently insectivorous. The Tree Swallow is known to feed on bayberries when its usual fare is wanting, but, with this exception, it is doubtful if any but insect food passes a Swallow's bill from one year's end to another. Recalling now the activity of Swallows, which both necessitates a large supply of food and procures it, and we must realize that these birds are incalculably beneficial.

Both the feeding habits and powers of flight of Swallows are such as their structure would lead us to expect, but when we examine their nests we are amazed at the architectural skill of builders so poorly provided with tools. The large mud pocket of the Barn Swallow, the clay retort of the Cliff Swallow, and the long burrow which the Bank Swallow excavates, are surely not the kind of homes we should expect these small-billed, weak-footed, dainty creatures to construct. We will note, too, that these feathered architects are quick to perceive and take advantage of the new and favorable conditions for nest-building found about the home of man.

The Bank, Rough-winged, and Tree Swallows, and the Purple Martin, lay white eggs; the eggs of the Barn and Cliff Swallows are speckled with cinnamon, olive, and reddish brown.

It is when nesting that Swallows best show one of their strong characteristics—their sociability. Many birds live in flocks during part of the year, but separate in pairs when nesting; but most Swallows live on terms of such intimacy that their nests seem to be merely apartments in one great dwelling. A photograph of part of a colony of Cliff Swallows in Montana shows one hundred and forty nests, nearly all of which adjoin one another.

The songs of Swallows are humble efforts, but are so expressive of the happy dispositions of the birds, and so associated with scenes with which they are inseparably connected, that the merry twitterings of these birds are as dear to us as the voices of friends.

The sociability of Swallows does not end with the nesting season, as it does with many birds that are then brought into communities by force of circumstances. When the young take wing, Swallows begin to collect in flocks, which gradually unite, and in August and September form assemblages containing millions of individuals. They generally make their headquarters in some large marsh, where they roost in the reeds and grasses, but they also resort to trees. Early in the morning they scatter over the country in small bands, flying at a considerable height, and during the day we may often see them feeding over fields and ponds or resting on wayside telegraph wires. Late in the afternoon they begin to return to their roosts. At first they fly slowly and circle about to feed, but as the light fails they fly with increasing swiftness, and the last comers shoot through the dusk with incredible rapidity.

These remarks apply with equal truth to all our Swallows; it remains now to briefly mention the characters by which they may be distinguished specifically. The four common species are figured in the frontispiece, which clearly shows most of their diagnostic marks, which are: Tail forked, Barn Swallow; forehead whitish, rump rusty, Cliff Swallow; a band across the breast, plumage without metallic colors, Bank Swallow; breast pure white, Tree Swallow.

The Barn Swallow is the most generally distributed of our Swallows, its habits of nesting in outbuildings making it at home wherever they offer it a suitable nesting place. It is about seven inches long; the upper parts and sides of the breast are steel-blue, the forehead and throat chestnut, the rest of the under parts paler; the tail deeply forked and marked with white. Its long tail is a most efficient rudder, permitting the abrupt turns which make its flight more erratic than that of any other of our Swallows. It skims low over the fields, or darts through the village streets with a rapidity and indirectness which I never witness without astonishment.

Barn Swallow,
Chelidon erythrogaster.
(Frontispiece.)

The Barn Swallow arrives from its winter home in the tropics about April 15 and remains until late in September. Its nest is generally placed on a beam in a barn or other outbuilding, and is composed of mud and grasses lined with feathers.

The Cliff or Eave Swallow is less generally distributed than the Barn Swallow. It nests in colonies, placing its rows of mud tenements under cliffs in the West and beneath the eaves of barns in the East. It becomes much attached to one locality, and when undisturbed returns to it year after year, arriving from the South about May 1, and remaining until late September. It is six inches long; the forehead is whitish, the crown and back steel-blue, the rump rusty; the throat chestnut with a blackish area; the belly white.

Cliff Swallow,
Petrochelidon lunifrons.
(Frontispiece.)

Like the Cliff Swallow, the Bank Swallow nests in colonies, and is very local during the breeding season. A sandbank facing a stream or pond is often chosen for a home. Into it a tunnel two or three feet in length is bored, and at its end a nest of grasses and feathers is built.

Bank Swallow,
Clivicola riparia.
(Frontispiece.)

The Bank Swallow winters in the tropics and reaches us in the spring about April 20, remaining until late September. It is the smallest of our Swallows, measuring only five inches in length, and is the only one, except the Rough-winged Swallow, which has no metallic coloring in its plumage, the back being plain brownish gray, the under parts white, with a clearly defined brownish gray band across the breast. The Rough-wing is a more southern bird, being rare north of southern Connecticut. It resembles the Bank Swallow, but differs chiefly in having the whole breast brownish gray. It nests in holes in banks, and also about stone bridges, trestles, and similar structures.

Though very generally distributed, there are large areas within the breeding range of the Tree Swallow **Tree Swallow,** where it is known only as a migrant. *Tachycineta bicolor.* In the wilder part of its range it nests (Frontispiece.) in hollow trees ; in the more settled portions it uses bird-boxes. During recent years, as Mr. Brewster has remarked, the always-present House Sparrow has pre-empted the former abodes of the Tree Swallow, so that it no longer nests about our homes ; but as a migrant its numbers are undiminished, and it is probably our most abundant Swallow.

Being the only Swallow to winter in the eastern United States, the Tree Swallow is the first to arrive in the spring, coming to us from Florida early in April. It is also the last of its family to leave us in the fall, often remaining near New York city until October 20.

Immature birds have the upper parts brownish gray instead of shining steel-blue, as in the adult, but in either plumage the bird may be known by its pure white under parts, which have given to it the name of White-bellied Swallow.

In the northern United States Martins are very local.

They have long since abandoned their habit of building in hollow trees, and now nest only about houses or in lawns where gourds or boxes are erected for their occupation. To these they return year after year, arriving in the spring about April 25 and remaining until September. The male is uniform steel-blue, and appears black in the air; the female is grayish, tinged with steel-blue above; the breast is gray, the belly white. This is the largest of our Swallows, measuring eight inches in length.

Purple Martin,
Progne subis.

WAXWINGS. (FAMILY AMPELIDÆ.)

One of the two species of Waxwing is a bird of the far North; the other, our Cedar Waxwing, is found throughout North America. Waxwings possess in an unusual degree two characteristics which are not supposed to be associated—sociability and silence. None of our birds is more companionable, none more quiet. In their fondness for one another's society they seem to delay the pairing season, and long after other birds have gone to housekeeping they are still roving about in flocks. Finally, late in June, they settle down and build a nest of generous proportions, often in some fruit tree, about ten feet from the ground. The three to five eggs are pale bluish gray or putty-color, spotted with black or brownish black.

Cedar Waxwing,
Ampelis cedrorum.
Plate LVII.

Waxwings fly in close rank and alight as near each other as the nature of their perch will allow. They sit very still, like little Parrots or Doves, but often raise and lower their crests, and perhaps whisper a fine lisping note, which is prolonged into a louder call—a string of beady notes—as they take wing.

Their fare varies with the season—cedar berries, strawberries, cherries, both cultivated and wild, the berries

CEDAR WAXWING.

Length, 7·20 inches. Grayish brown ; upper tail-coverts gray ; lower belly
yellowish ; end of tail yellow ; secondaries sometimes with red, sealing-wax-
like tips ; stripe through face black.

of the woodbine, sour gum, and others being taken in turn.

In August the Waxwing shows no mean gifts as a flycatcher, while as a destroyer of the cankerworm he is especially beneficial, repaying us with interest for the fruit he may have appropriated earlier in the season.

The Waxwing's wide range and ability to withstand great extremes in temperature are doubtless due to the ease with which it adapts itself to a change in fare. It nests from Virginia to Labrador, and winters from Massachusetts to Costa Rica.

SHRIKES. (FAMILY LANIIDÆ.)

The marked difference in the temperament of birds is emphasized by finding among the song birds, who feed

Northern Shrike,
Lanius borealis.
Plate LVIII.

on fruit, seeds, and insects, a bird who in his position and choice of food is truly hawklike. Shrikes are solitary, never assembling in flocks or associating with other birds. Their days are days of waiting, varied by a pounce upon some unfortunate field mouse or dash into a flock of unsuspecting Sparrows. But, while they resemble the Hawks in these respects, their manner of capturing their prey differs from that of their larger prototypes. The Shrike has a Hawk's bill but a Sparrow's foot, and, lacking the powerful talons which make so deadly a weapon, he captures his prey with his strong mandibles. Possibly it may be due to his comparatively weak feet that he pursues the singular custom of impaling his prey on some thorn or hanging it from a crotch where he can better dissect it.

The Shrike, or Butcher-bird, as he is also called, belongs to a large family, but, with the exception of his smaller cousin the Loggerhead, he is the only one of the two hundred known species found in America. He nests

PLATE LVIII.

NORTHERN SHRIKE.

Length, 10·30 i ches. *Adult*, upper parts gray ; tail black and white ; under parts white, with blackish bars ; lores grayish ; ear-coverts black. *Young*, similar, but plumage washed with brownish.

within the Arctic Circle, and in October journeys southward, rarely as far as Virginia, and remains in the United
States until April or May.

The Loggerhead Shrike is common in the Southern
States and Mississippi Valley, whence it has apparently
extended eastward through central New York to Vermont and Maine. It nests in these States, but southward
to Maryland is known only as a rare migrant—a unique
case in distribution. It differs from the Northern Shrike
in being an inch and a quarter smaller, in the absence of the
wavy bars on the breast, which is pure white, and in having jet-black lores and a narrow black line across the forehead at the base of the bill. Its song is *creaky* and unmusical, but the song of the Northern Shrike, as described by Mr. Brewster, is " really pleasing," and " not
unlike that of the Thrasher, but more disconnected and
less loud and varied."

VĬREOS. (FAMILY VĬREONIDÆ.)

Vireos are gleaners, and are to be distinguished from
other tree-inhabiting, greenish birds of the same size by
their habit of carefully exploring the under surface of
leaves and various nooks and corners in the bark and
foliage, while the more active Warblers are flitting about
the terminal twigs and the Flycatchers are swinging out
in aërial loops at passing insects.

They are highly musical little birds, having songs and
call-notes which may be quickly recognized once they
are known. The nests and eggs of our four summer-
resident species are so much alike that they are to be
known only when accompanied by their owners. The
White-eyed Vireo inhabits thickets and, as a rule, builds
nearer the ground than the arboreal Red-eyed, Yellow-
throated, and Warbling Vireos. The nests are small,

pouchlike affairs of strips of pliable bark, bits of dead wood, plant-fibers, tendrils, fine grasses, etc., firmly interwoven and suspended from the arms of a forked twig. The eggs are white, with a few black or brownish black spots, chiefly about the larger end.

The Vireos are an exclusively American family, and number some fifty species, of which seven reach the northeastern States. Of these, by far the most common is the Red-eyed Vireo. There are few favorable localities in eastern North America where, in the summer, one may not hear the cheerful song of this bird. Still, it is so well protected by the foliage, with which its plumage agrees in color, that to those· whose ear is not attuned to the music of birds it is unknown. But listen near some grove of elms or maples, and you will not fail to hear its song—a somewhat broken, rambling recitative, which no one has described so well as Wilson Flagg, who calls this bird the Preacher, and interprets its notes as " You see it—you know it—do you hear me ?—do you believe it ? " The Red-eye evidently has an inquiring mind, for he never tires of asking these questions. He not only sings all day, but seems unaffected by the heat of summer, and at midday is often the only bird to be heard. One would imagine that few birds had a more even temperament than this calm-voiced singer, but when annoyed he utters a complaining *whang*—a sound which is a good indication that something is wrong in the bird world.

Red-eyed Vireo,
Vireo olivaceus.
Plate LIX.

The Red-eye winters in the tropics, and reaches us in the spring about May 1, remaining until October 15.

A near relative of the Red-eye's is the Warbling Vireo—a somewhat smaller bird, with a brown, in place of red eye, and without the black margin above the white eye-line which can be so easily seen in the Red-eye. The Warbling Vireo is the less common of the two, and is

more local, showing a marked fondness for rows of elms
—a taste which makes it a dweller in towns and villages.

Its song bears no resemblance to that
Warbling Vireo,
Vireo gilvus.
of the Red-eye, being a continuous,
flowing warble, with an alto under-
tone, suggestive of the song of the Purple Finch.

The Warbling Vireo arrives from its winter home in
the tropics about May 5, and remains until late in September.

Although the Yellow-throated Vireo is least like the
Red-eye in color, it resembles it the most closely in choice
Yellow-throated of haunts and in song. Still, the Yellow-
Vireo, throat's song is sung more deliberately
Vireo flavifrons. and with longer pauses between the
Plate LIX. parts, while in tone it is deeper and
richer. To my mind he says : " See me ; I'm here ; where
are you?" repeating the question in varying forms.
Rarely he utters a beautiful, mellow trill which suggests
the song of the Ruby-crowned Kinglet, and he has also
a *cacking*, scolding note like that of the White-eye.
The Yellow-throat's nest is often a more elaborate structure than those of our other Vireos, being thickly covered with lichens, which add greatly to its beauty.

Like the two preceding species, the Yellow-throat
winters in the tropics, and reaches the latitude of New
York city about May 1. It does not, however, remain
as long as its relatives, leaving us about September 15.

The White-eyed Vireo is the genius of his family.
What the Chat is among Warblers the White-eye is
among Vireos—a peculiar, eccentric
White-eyed Vireo,
Vireo noveboracensis.
bird of strong character, who regards
mankind with disapproval, and will
have none of us. Excellent reasons these why we
should court his acquaintance.

Unlike our other Vireos, the White-eye lives in the

PLATE LIX.

RED-EYED VIREO.

Length, 6·25 inches. Crown gray, bordered by black and white ; back, wings, and tail olive-green ; under parts, white.

YELLOW-THROATED VIREO.

Length, 5·95 inches. Crown and back greenish yellow ; rump gray ; breast bright yellow ; belly white ; wing-bars white.

lower growth; thickets of cat-brier are his favorite haunts. He is therefore nearer our level, and seems to address us more directly than do the birds that call from the tree tops. If you linger near his home he will inquire your business with a vigorous " I say, who are you, eh ? " and if you do not take this hint to move on he will doubtless follow it with a scolding whose intent is unmistakable. He has a variety of exclamatory calls, and sometimes may be heard softly singing a song composed largely of imitations of the notes of other birds.

The White-eye can easily be known from the Red-eye and Warbling Vireos by the narrow white bands across the tips of its wing-coverts. In this respect it resembles the Yellow-throat, from which it is to be distinguished by its smaller size (length 5·25 inches), white iris, and white breast, only the sides of the breast being tinged with yellow. It winters from Florida southward, and reaches us in the spring about May 1, to remain until October.

WARBLERS. (FAMILY MNIOTILTIDÆ.)

Warblers may be described as among our most abundant, most beautiful, and least-known birds. Of the thirty-five species regularly found in the northeastern States, only three or four are familiar to the casual observer. The presence of the others is unsuspected, and when some chance brings one of these exquisite little creatures into our lives, the event is attended by all the excitement of an actual discovery. We never forget our first Warbler.

It is because we do not see Warblers unless we look for them that they are strangers to so many persons who go to the woods. They are, with some exceptions, small birds of limited vocal powers. They live in the tree tops,

Creeping Lichen

PLATE LX.

BLACK AND WHITE WARBLER.

Length, 5 25 inches. *Adult male*, upper parts, breast, and sides black and white ; belly white. *Adult female*, similar, but with less black on under parts.

16

and their lisping notes blend with other woodland voices without attracting our attention.

May and September are the months for Warblers. Some species arrive in April, but they are most numerous between May 5 and 15, when the woods are thronged with their flitting forms. Less than half of our thirty-five species remain to breed ; the others go to their summer homes in the coniferous forests of the North. These northern birds return in the latter part of August and abound in September. Many of the Warblers seen at this season are immature birds wearing plumages so different from those of the adult birds seen in the spring, that their identity is not suspected, and, in effect, they are new birds to us.

To the field ornithologist Warblers are therefore the most difficult as well as the most fascinating birds to study. Long after the Sparrows, Flycatchers, and Vireos have been mastered, there will be unsolved problems among the Warblers. Some rare species will be left to look for—it may be a member of the band flitting about actively in the branches above us—and in the hope of finding it we eagerly examine bird after bird until our enthusiasm yields to an aching neck.

Acquaintance with more familiar birds will doubtless arouse the enthusiasm necessary to a successful pursuit of

Black and White Warbler, *Mniotilta varia.* Plate LX.

Warblers, but in the meanwhile I will mention only those species that can be most easily observed. Among them is the Black and White Warbler, whose habit of creeping or climbing over trunk and limb aids in his identification. He is a summer resident, and about April 20 we may expect to hear the thin, wiry *see-see-see-see* notes which form his song. A month later we may find his nest, placed on the ground at the base of a stump or stone and containing four or five white

MYRTLE WARBLER.

Length, 5·65 inches. *Winter plumage,* crown-patch, rump, and sides of breast yellow ; back brown and black ; under parts black and white. *Summer plumage,* similar, but upper parts gray and black ; more black on under parts.

BLACK-THROATED GREEN WARBLER.

Length, 5 10 inches. Upper parts yellowish green ; face brighter ; breast black ; belly white.

eggs speckled with reddish brown, chiefly at the larger end.

The Yellow Warbler is also a summer resident, arriving in the spring about April 30 and remaining, with the Black-and-white Warbler, until late in September. It has the general appearance of being an entirely yellow bird, and is sometimes called " Wild Canary," but it has a much more slender bill than the Canary, and its breast is spotted with reddish brown. Most Warblers are wood-inhabiting birds, but the Yellow Warbler, unlike its relatives, prefers lawns, parks, and orchards to woodlands. Its nest, of fine grasses, fibers, and a large amount of cottony plant-down, is placed in shrubbery or shade trees. Its eggs are bluish white, thickly marked with cinnamon and olive-brown.

Yellow Warbler,
Dendroica æstiva.

The Black-throated Green Warbler nests in pine forests from southern New England northward, arriving from the South about May 1 and remaining until October. Its nest is usually placed in pine trees; its eggs are white, spotted and speckled with dark brown.

Black-throated
Green Warbler,
Dendroica virens.
Plate LXI.

The songs of many Warblers are possessed of so little character that the best description conveys no idea of them, but the quaint *zee-zee, zee-ee, zee* of the Black-throated Green, which Mr. Burroughs writes — — ᴠ —, will be readily recognized.

The Myrtle or Yellow-rumped Warbler nests from northern New England northward, and in winter is the only Warbler to remain in the Northern States, being often found as far north as New York city, when its favorite food of bayberries can be procured. At this season there is little or no black on the breast and the

Myrtle Warbler,
Dendroica coronata.
Plate LXI.

The Torch-bearer

Ernest Seton Thompson

PLATE LXII.

REDSTART.

Length, 5·40 inches. *Adult male*, band in wings ; base of tail and sides of breast deep salmon ; belly white ; rest of plumage black. *Adult female and young*, similar, but salmon replaced by yellow ; upper parts grayish brown ; under parts white ; breast yellowish.

back is grayish brown, but this Warbler may always be known by its four patches of yellow and its characteristic call-note of *tchip*.

The Redstart belongs to the group of fly-catching Warblers, and, as an indication of its manner of feeding,

Redstart,
Setophaga ruticilla.
Plate LXII.

his bill is much broader and flatter than is usual in this family. The Redstart is not so patient and methodic a flycatcher as the birds to whom this name rightly belongs. They sit quietly until some insect comes within reach, and then with unerring aim launch out at it, returning to their perch to devour it at leisure. But the Redstart darts here and there, falls and rises and spins about, catching an insect at every turn and at the same time displaying his bright colors to such advantage that he seems the most beautiful as he is the most animated bird of the woods. As he pirouettes from limb to limb, with drooped wings and spread tail, he sings *ser-wee swee, swee-ee*, a simple but merry little jingle.

The Redstart's bright colors, like some mark of special distinction, are not acquired at once. The young male must pass through a period of probation before he is worthy to wear the orange-red and black. In the meantime he appears first in the costume of the female, and by successive changes reaches the full dignity of Redstart estate at the age of three years. He nests, however, the first year, when his plumage closely resembles that of his mate. The nest, of fine strips of bark, plant-down, and other materials, is built in the crotch of a sapling ten to twenty feet from the ground. The eggs are grayish white or bluish white, spotted and blotched, chiefly at the larger end, with cinnamon and olive-brown. They are laid about May 28—four weeks after the bird's arrival from the South.

All the Warblers thus far mentioned are tree-inhabit-

Plate LXIII.

OVEN-BIRD.

Length, 6·15 inches. Crown reddish brown, bordered by black ; back, wings, and tail olive-green ; under parts black and white.

231

ing birds, but the species now to be spoken of pass most of their time in the undergrowth or on the ground. The

Oven-bird,
Seiurus aurocapillus.
Plate LXIII.

Oven-bird chooses the latter locality. He has been well compared by Mr. Burroughs to a little Partridge, and if you have enough perseverance to find the author of the sharp *cheep* with which this somewhat suspicious bird will greet you, you will see a modestly attired little walker daintily picking his way over the leaves and fallen branches, with crest slightly erect, and head nodding at each step.

Probably, however, your first acquaintance with the Oven-bird will be made through the medium of his song. There are few bits of woodland where in May and June you can not hear numbers of these birds singing. It is a loud, ringing, crescendo chant, to which Mr. Burroughs's description of "teacher, *teacher*, TEACHER, TEACHER, **TEACHER**" is so applicable that no one would think of describing it in any other way. The bird seems to exert himself to the utmost, and no one hearing this far from musical performance would imagine that he could improve upon it. But if some evening during the height of the mating season you will visit the Oven-bird's haunts, you may hear a song whose wildness is startling. It is the flight-song of the Oven-bird, transforming the humble chanter into an inspired musician. Soaring high above the trees, he gives utterance to a rapid, ecstatic warbling so unlike his ordinary song that it is difficult to believe one bird is the author of them both.

As an architect the Oven-bird is also distinguished. His unique nest is built on the ground of coarse grasses, weed stalks, leaves, and rootlets, and is roofed over, the entrance being at one side. It thus resembles an old-fashioned Dutch oven, and its shape is the origin of its builder's name. The Oven-bird arrives from the South

MARYLAND YELLOW-THROAT.

Length, 5·30 inches. *Adult male*, face black, bordered by ashy ; back olive-green ; breast yellow ; belly paler. *Adult female*, similar, but no black on face ; under parts paler.

about May 1, and its eggs are laid about the 20th of the month. They are white, speckled or spotted with cinnamon and reddish brown.

The Maryland Yellow-throat is an abundant inhabitant of thickets and bushy undergrowths, readily identified by his black mask and yellow breast, nervous activity, and characteristic notes. Some birds must be approached with caution, but nothing save an actual attack upon his home will cause the Yellow-throat to leave its shelter. Hopping from limb to limb, he advances to the border of the thicket, then retreats to its depths, all the time uttering an impatient *chack, chit,* or *pit,* and, if forced to fly, he goes only to the next clump of bushes.

Maryland Yellow-throat,
Geothlypis trichas.
Plate LXIV.

The Yellow-throat's somewhat explosive song is so easily set to words and so variable that there are many versions of it. It is described as *whitititee, whitititee, whitititee; rapity, rapity, rapity, rap,* or *witch-e-wee-o, witch-e-wee-o, witch-e-wee-o.* Mr. Burroughs says he has heard birds whose notes sounded like the words " Which way, sir?" and I have heard some who seemed to say " Wait a minute."

To this the Yellow-throat sometimes adds a flight song, which is a miniature of the Oven-bird's aërial serenade. It is generally added to his usual song, and is most often heard late in the season at evening, when the bird may be seen springing into the air above his bushy retreat.

The Yellow-throat arrives from the South about May 1, and remains until the middle of October. Late in May a bulky nest of grasses, strips of bark, and dead leaves, lined with finer materials, is built on or near the ground. The three to five eggs are white, rather thinly speckled with reddish brown. Often an egg of the Cow-

PLATE LXV.

YELLOW-BREASTED CHAT.

Length, 7·45 inches. Upper parts olive-green ; breast yellow ; belly white ;
lores black, bordered by white.

bird will be found in the nest, Yellow-throats being one of the birds most frequently chosen by the Cowbird as foster-parents.

The Chat is the largest of the Warblers, and so unlike them, or any other birds, in disposition that if classifica-
Yellow-breasted Chat, tion were based on character, the Chat
Icteria virens. would surely be placed in a family by
Plate LXV. itself. The Chat's peculiarities are numerous, but are most evident in his song. Many times I have sat, note-book and pencil in hand, trying to express in words the song of a Chat singing in a neighboring thicket, but I have never succeeded in putting on paper anything which would convey an adequate idea of the bird's remarkable vocal performances. Of others who have attempted the same task, I think Mr. Burroughs comes nearest to interpreting the bird's strange medley. He says: "Now he barks like a puppy, then quacks like a Duck, then rattles like a Kingfisher, then squalls like a fox, then caws like a Crow, then mews like a cat. . . . *C-r-r-r-r-r—whrr—that's it—chee—quack, cluck, yit-yit-yit—now hit it—tr-r-r-r—when—caw—caw —cut, cut—tea-boy—who, who—mew, mew.*" You may be pardoned for doubting that a bird can produce so strange a series of noises, but if you will go to the Chat's haunts in thickety openings in the woods, or other bushy places, and let him speak for himself, you will admit that our alphabet can not do him justice. To hear the Chat is one thing, to see him quite another. But he will repay study, and if you will conceal yourself near his home you may see him deliver part of his repertoire while on the wing, with legs dangling, wings and tail flapping, and his whole appearance suggesting that of a bird who has had an unfortunate encounter with a charge of shot.

But if the Chat's song is surprising when heard during the day, imagine the effect it creates at night when he has the stage to himself, for he is one of our few birds who sing regularly and freely during the night, moonlit nights being most often selected.

The Chat is a rather southern bird in its distribution, being found north of Connecticut only locally and rarely. It winters in the tropics, coming to us about May 1 and departing early in September. Its well-made nest of grasses, leaves, and strips of bark is generally placed in the crotch of a sapling within three feet of the ground. Its three to five eggs are white, rather evenly speckled and spotted with reddish brown.

THRASHERS, WRENS, ETC. (FAMILY TROGLODYTIDÆ.)

The Eastern representatives of this family are apparently too unlike to be classed in the same group, but when all the two hundred members of the family are studied, it is evident that the extremes are connected by intermediate species possessing in a degree the characters of both Wrens and Thrashers.

The Catbird belongs to the subfamily *Miminæ*, which contains also the Mockingbirds and Thrashers, number-

Catbird,
Galeoscoptes carolinensis.

ing some fifty species, all being restricted to North America. The Catbird is one of the most familiar feathered inhabitants of the denser shrubbery about our lawns and gardens. The sexes are alike in color, both being slaty gray, with a black cap and tail, and brick-red under tail-coverts. They arrive from the South about April 29, and remain until October. Their nest is usually placed in thickets, shrubbery, or heavily foliaged trees, and the deep greenish blue eggs are laid the fourth week in May.

It is unfortunate that the Catbird's name should have originated in his call-note rather than in his song. The

former is a petulant, whining, nasal *tchay*, to me one of
the most disagreeable sounds in Nature, and so unlike
the bird's song that he seems possessed of a dual person-
ality. The Catbird's song, from a musical standpoint, is
excelled by that of few of our birds. His voice is full
and rich, his execution and phrasing are faultless; but
the effect of his song, sweet and varied as it is, is marred
by the singer's too evident consciousness.

The Catbird's relative, the Mockingbird, is an abundant inhabitant of the southern United States from Virginia to California, and ranges south-
ward into Mexico. In the Eastern
States it is not common north of south-
ern Illinois and Virginia, but in summer it is found in
small numbers as far north as Massachusetts, where a few
pairs breed each season. It is exceedingly domestic in
its habits, and in the South there are few suitable gardens, either in the town or country, which are not inhabited by a pair of Mockingbirds.

Mockingbird,
Mimus polyglottos.
Plate LXVI.

The power of mimicry for which this bird is cele-
brated has, I think, been unduly exaggerated, and the
fact that its usual song contains several notes resembling
those of other species doubtless in part accounts for its
much overrated ability as a mimic. It is unnecessary,
however, for the Mockingbird to borrow the notes of other
birds, for his own song places him in the front rank of
our songsters. It is delivered with a spirit and animation
which add greatly to its attractiveness. The Mockingbird does not sing between mouthfuls, as do the Vireos,
or quietly from a perch, like the Towhee or Thrasher;
he frequently changes his position, hopping from place
to place, making short flights, bounding into the air, and
displaying the white markings of his wings and tail, as
though it were impossible for him to give expression to
his emotion through the medium of voice alone. During

MOCKINGBIRD.

Length, 10 50 inches.   Upper parts ashy gray ; wings and tail brownish black and white ; under parts white.

moonlight nights of the nesting season, Mockingbirds sing all night. They are then less active, and, mounting to some favorite perch, often a chimney top, flood the still air with entrancing melody.

Like the Catbird and Mocker, the Brown Thrasher or Brown "Thrush" inhabits thickets and undergrowth.

Brown Thrasher, He is, however, a much less domestic
Harporhynchus rufus. bird, and prefers brushy pasture lots and
Plate LXVII. wayside hedges to lawns or gardens.
He arrives from the South the latter part of April, and often remains until late in October. The nest is built about May 15, and is placed on the ground or several feet above it. The eggs are bluish or grayish white, thickly, evenly, and minutely speckled with cinnamon or reddish brown.

As a songster I should rank the Thrasher between the Mocker and the Catbird. His song is less varied and animated than the Mocker's, and while his technique may not excel that of the Catbird, his song, to my mind, is much more effective than the performance of that accomplished musician. Mounting to the topmost limb of a tree, he sings uninterruptedly for several minutes. The notes can be heard for at least a third of a mile, ringing out clear and well defined above the medley of voices that form the chorus of a May morning.

The intense vitality which characterizes the life of birds finds its highest expression in the Wrens. Perpet-

House Wren, ual motion alone describes the activity
Troglodytes aëdon. of these nervous, excitable little crea-
Plate LXVIII. tures. Repose seems out of the ques-
tion; as well expect to catch a weasel asleep as to find a Wren at rest.

In his movements, song, and nesting habits our House Wren exhibits the characteristic traits of his family. He is ever hopping, flitting, bobbing, or bowing, pausing

PLATE LXVII.

BROWN THRASHER.

Length, 11.40 inches. Upper parts bright reddish brown ; under parts white and black ; eyes yellow.

17

only long enough to give voice to his feelings in fidgetty, scolding notes, or an effervescing, musical trill, with the force of which his small body trembles. It is a wonderful outburst of song, and the diminutive singer's enthusiasm and endurance are even more remarkable. The song occupies about three seconds, and I have heard a Wren, in response to a rival, sing at the rate of ten songs a minute for two hours at a time.

The House Wren nests in almost any kind of suitable hole or cavity, and will frequently take possession of a bird box, if the House Sparrows have not already set up a claim to the same property. To prevent intrusion from the Sparrows, the entrance to the house should be made not larger than a quarter of a dollar. Whatever be the site the Wrens select, their surplus energy is employed in completely filling it with twigs, half a bushelful being sometimes brought with endless pains. The nest proper is composed of dried grasses, and is placed in the center of this mass. Even in egg-laying the exhaustless vitality of Wrens is shown, as many as six or eight eggs being deposited. In color they are uniformly and minutely speckled with pinkish brown.

The House Wren arrives from the South late in April and remains until October. Shortly before its departure in the fall a Wren comes from the **Winter Wren,** *Troglodytes hiemalis.* North that resembles the House Wren in appearance, but is smaller and has the under parts pale brown, the breast and belly being finely barred with a darker shade of the same color. This is the Winter Wren, a bird that nests from northern New England northward and southward along the crests of the Alleghanies to North Carolina. It remains with us in small numbers throughout the winter, returning to its summer home in April. Mr. Burroughs writes of the Winter Wren's song as a " wild, sweet, rhythmical

Plate LXVIII.

HOUSE WREN.

Length, 5·00 inches. Upper parts brown, marked with black and grayish; under parts grayish white.

cadence that holds you entranced," but while with us the bird's only note is an impatient *chimp, chimp*, suggesting the Song Sparrow's call-note.

The Carolina Wren is a more southern bird than the House Wren. It is of only local distribution north of **Carolina Wren,** southern New Jersey, and is rarely *Thryothorus* found north of the vicinity of New *ludovicianus.* York city, where it appears to be increasing in numbers and is found throughout the year. This Wren is half an inch longer and decidedly heavier than the House Wren; its upper parts are bright cinnamon, its under parts washed with the same color, and a conspicuous white line passes from the bill over the eye.

The Carolina Wren is an exceedingly musical bird, and its loud whistled calls are among the most characteristic bird notes in the South. They are numerous and varied, the most common resembling the syllables *whee-udel, whee-udel, whee-udel*, and *tea-kettle, tea-kettle, tea-kettle.*

The haunts of most marsh-inhabiting birds are as sharply defined as the limits of their ranges. The Long-**Long-billed** billed Marsh Wren is not known in **Marsh Wren,** the East north of Massachusetts, but I *Cistothorus palustris.* would as soon expect to find one of Plate LXIX. these birds in Greenland as out of a marsh. They arrive from the South early in May and remain until October, living in marshes where cat-tails grow, to which they may attach their bulky, globular nests of reeds and grasses. With the superabundant vigor of Wrens they build more nests than they can possibly occupy, and many will be examined before the five to six dark brown eggs are found.

The Marsh Wren is quite as active and irrepressible as the other members of his family. His call is the cus-

PLATE LXIX.

LONG-BILLED MARSH WREN.

Length, 5·20 inches. Upper parts brown, black, and white, a white line over eye ; under parts white, sides brownish.

245

tomary scolding *cack;* his song, a bubbling, trickling tinkle that can not be called musical, but to my mind is indescribably attractive. It is often sung in the air, and in marshes where Wrens are abundant bird after bird may be seen springing a few feet above the reeds, singing his song, and then dropping back again.

CREEPERS. (FAMILY CERTHIIDÆ.)

Of the twelve known members of this family, the Brown Creeper is the only one inhabiting the New World. It is a northern bird, breed-

Brown Creeper,
Certhia familiaris americana.
Plate LXX.

ing at sea level only from Maine northward, but extending southward in the Alleghanies to North Carolina. Several western races are found in the Rocky Mountain region and Sierra Madres. Our eastern bird migrates southward late in September, and from that date until April it may be found from Massachusetts to Florida.

The Creeper, like a Woodpecker, never climbs head downward, but, using his stiff, pointed tail-feathers (see Fig. 3 *b*) as a prop, winds rapidly up the trunks of trees in his apparently never-ending search for insects' eggs and larvæ hidden in crevices in the bark. If the Wrens are the most active birds, the Creeper is the most diligent. Except when it was stopping to secure some tidbit, I can not remember seeing a Creeper resting. He usually begins at the base of a tree and climbs in a serious, intent way for a certain distance, and then, without a moment's pause, drops down to the bottom of the next tree and continues his search.

The Creeper's only notes while with us are a thin, fine squeak; but Mr. Brewster tells us that during the nesting season he has an exquisitely tender song of four notes.

CHICKADEE.

Length, 5.25 inches. Crown and throat black ; cheeks white ; back gray ; belly white, washed with brownish.

BROWN CREEPER.

Length, 5.65 inches. Upper parts brown, rusty, and white ; under parts white.

TITMICE AND NUTHATCHES. (FAMILY PARIDÆ.)

Comparing the Titmice with the Nuthatches, the former may be described as short-billed birds with long tails who do not creep, the latter as long-billed birds with short tails who do creep. The two groups are, in fact, quite distinct, and by some systematists are placed in separate families.

The Titmice number some seventy-five species, four of which are found in eastern North America. The

Chickadee, *Parus atricapillus.* Plate LXX.

commonest and most generally distributed is the Black-capped Chickadee, which is found from Labrador to Maryland and in the Alleghanies southward to North Carolina. Farther south it is replaced by the closely allied Carolina Chickadee.

Throughout the greater part of its range the Chickadee is found at all seasons, but it is less common in the middle and southern New England States in summer than in winter, and is most numerous during its migration in October.

It is with winter that these merry little black and white midgets are generally associated. Their tameness, quaint notes, and friendly ways make them unusually companionable birds; one need not lack for society when Chickadees are to be found. Many of their notes are especially conversational in character, and in addition to the familiar *chickadee* call, they have a high, sweet, plaintive two- or three-noted whistle.

The Chickadee nests about the middle of May, selecting some suitable cavity or making one for himself in a decayed trunk or limb and lining it with moss, plant-down, and feathers. The eggs, five to eight in number, are white, spotted and speckled, chiefly at the larger end, with cinnamon or reddish brown.

PLATE LXXI.

RED-BREASTED NUTHATCH.

Length, 4 60 inches. *Male*, crown and line through eye black ; back gray ;
under parts rusty. *Female*, similar, but black replaced by gray.

WHITE-BREASTED NUTHATCH.

Length, 6·05 inches. *Male*, crown black ; back gray ; face and under parts
white. *Female*, similar, but crown slaty.

The Tufted Titmouse is a more southern bird than the Chickadee and is rarely found north of northern New Jersey, where, however, it remains throughout the year. It is six inches in length, gray above, whitish below, with a black forehead, reddish brown sides, and a conspicuous crest. Its usual call is a whistled *peto, peto, peto*, which it will utter for hours at a time. It has also a *de-de-de-de* call, suggesting the Chickadee's well-known notes, but louder and more nasal.

Tufted Titmouse,
Parus bicolor.

With no especial structure other than slightly enlarged toe nails, the Nuthatches still differ markedly from other birds in the ease with which they run up or down tree trunks. The tail is short and square and is not used in climbing. The bill is rather slender, but proves an effective instrument in removing insects' eggs and larvæ from crevices in the bark and even in excavating a nesting hole in some decayed limb. Several species also use it to crack or " hatch " nuts after they have wedged them in a convenient crevice.

White-breasted Nuthatch,
Sitta carolinensis.
Plate LXXI.

Of the three species of Nuthatches found in eastern North America the White-breasted is the most common and generally distributed, being a permanent resident from Florida to northern New England. Like many resident birds, it nests early, the five to eight eggs being deposited about April 20. They are white, thickly and evenly spotted and speckled with reddish brown and lavender.

This Nuthatch's usual call-note is a loud *yank, yank*, while its song is a singular, tenor *hah-hah-hah-hah-hah*.

The Red-breasted Nuthatch is a more northern bird than its larger, white-breasted cousin. At sea level it nests from Maine northward, but in the higher parts of

the Alleghanies it breeds as far southward as North Carolina. It comes to us from the north early in Septem-

Red-breasted
    Nuthatch,
Sitta canadensis.
Plate LXXI.

ber, and in the winter may be found in
varying numbers from Massachusetts to
the Gulf States. Its call-note is
higher, thinner, and· more drawled

than the vigorous *yank*, *yank* of the White-breasted
Nuthatch, and suggests the sound produced by a penny
trumpet.

KINGLETS, GNATCATCHERS, ETC. (FAMILY SYLVIIDÆ.)

Of the three subfamilies included in this family
we have representatives in eastern North America of
only two—the two Kinglets of the subfamily *Regu-
linæ* and the Blue-gray Gnatcatcher of the subfamily
Polioptilinæ. The Gnatcatcher is a southern bird, occurring only locally or as a straggler north of Maryland.
The Kinglets are both more northern in their distribution.

The Golden-crowned Kinglet nests from the northern tier of States northward and southward along the

Golden-crowned
    Kinglet,
Regulus satrapa.
Plate LXXII.

crests of the Alleghanies to North Carolina. In its autumnal migration it
reaches the vicinity of New York city
about September 20, and during the

winter may be found in varying numbers from Maine to
Florida.

The Golden-crown flits about the terminal twigs in its
search for insect food and reminds one somewhat of the
smaller, tree-inhabiting Warblers in habits. Its call is a
fine *ti*, *ti*, one of the highest and least noticeable notes
uttered by birds. Its song, which is rarely heard except
in its nesting range, is described by Mr. Brewster as beginning with a succession of five or six fine shrill, high-pitched,

somewhat faltering notes, and ending with a short, rapid, rather explosive warble.

The Ruby-crowned Kinglet is a more northern bird in summer and a more southern bird in winter than the Golden-crown, rarely being found at the latter season north of South Carolina. Throughout the Middle States it oc-

Ruby-crowned Kinglet,
Regulus calendula.
Plate LXXII.

curs as an abundant spring and fall migrant, passing northward from April 10 to May 10 and southward between September 20 and October 20. The Ruby-crown resembles the Golden-crown in habits, but is more active. Females and young males lack the ruby crown-patch, but their white eye-ring, impatient, wrenlike little note, and manner of nervously twitching their wings are characteristic.

Taking the small size of the bird into consideration, the Ruby-crown's song is one of the most marvelous vocal performances among birds. As Dr. Coues remarks, the sound-producing organ is not larger than a pinhead, and the muscles that move it are almost microscopic shreds of flesh; still, the bird's song is not only surpassingly sweet, varied, and sustained, but is possessed of sufficient volume to be heard at a distance of two hundred yards. Fortunately, the Ruby-crown sings both on its spring and fall migrations.

THRUSHES, BLUEBIRDS, ETC. (FAMILY TURDIDÆ.)

On the basis of certain details of structure Thrushes are generally assigned highest rank in the class Aves. Without pausing to discuss the value of the characters on which this classification is made, there can be no question that from an æsthetic standpoint the Thrushes possess in a greater degree than any other birds those qualifications which make the ideal bird. There are many birds with

GOLDEN-CROWNED KINGLET.

Length, 4 05 inches. *Male*, crown orange, yellow. and black ; back olive-green ; under parts whitish. *Female*, similar, but crown without orange.

RUBY-CROWNED KINGLET.

Length, 4·40 inches. *Adult male*, crown-patch ruby ; back olive-green ; under parts whitish. *Adult female and young*, similar, but no crown-patch.

brighter plumage, more striking voices, and more interesting habits, but there are none whose bearing is more distinguished, whose songs are more spiritual. The brilliant
Hummingbirds and Tanagers excite our admiration, but
the gentle, retiring Thrushes appeal to our higher emotions; their music gives voice to our noblest aspirations.

Five of the true Thrushes of the genus *Turdus* are
found in eastern North America. Three of them may
be mentioned here—the Veery, Wood Thrush, and Hermit Thrush—a peerless trio of songsters. The Veery's
mysterious voice vibrates through the air in pulsating
circles of song, like the strains of an Æolian harp. The
Wood Thrush's notes are ringing and bell-like; he sounds
the matin and vesper chimes of day, while the Hermit's
hymn echoes through the woods like the swelling tones
of an organ in some vast cathedral.

But it is impossible to so describe these songs that
their charm will be understood. Fortunately, all three
birds are abundant, and a brief account of their haunts
and habits will enable any one to find them.

The Veery, or Wilson's Thrush, winters in Central
America, and nests from northern Illinois and northern

Veery,   New Jersey northward to Manitoba and
Turdus fuscescens. Newfoundland and southward along the
Plate LXXIII.   Alleghanies to North Carolina. It comes
to us in the spring, about May 1, and remains until September 15. Near the middle of May it begins to build its
nest, placing it on or near the ground. Its eggs are
greenish blue, and resemble in color those of the Wood
Thrush, but are slightly smaller.

The Veery's favorite haunts are low, damp woods
with an abundant undergrowth. It is a more retiring
bird than the Wood Thrush, and is rarely seen far from
tracts of woodland. It is to be distinguished from our
other Thrushes by the uniform cinnamon color of its

PLATE LXXIII.

VEERY.

Length, 7·50 inches. Upper parts, wings, and tail uniform light cinnamon ; breast buffy, light marked with cinnamon ; belly white ; sides grayish.

255

upper parts, faintly spotted breast, and particularly by its notes.

The Veery's characteristic calls are a clearly whistled *whèe-o* or *whèe-you*, the first note the higher, and a somewhat softer *tòo-whee* or *teweù*, in which the first note is the lower. Its song is one of the most mysterious and thrilling sounds to be heard in the woods. Elsewhere-I have described it as "a weird, ringing monotone of blended alto and soprano tones. . . . It has neither break nor pause, and seems to emanate from no one place. If you can imagine the syllables vee-r-r-hu [or vee-ry] repeated eight or nine times around a series of intertwining circles, the description may enable you to recognize the Veery's song."

The Wood Thrush is a more southern bird than the Veery, breeding from as far south as Florida, north-

Wood Thrush, ward to southern Vermont and Minne-
Turdus mustelinus. sota. It winters in Central America
Plate LXXIV. and reaches us in the spring, about April 30, and remains until October 1. Its nest is built about the middle of May, and is generally placed in a sapling some eight feet from the ground. The eggs are greenish blue.

The Wood Thrush is not such a recluse as the Veery. He is, it is true, a wood lover, and shares with the Veery his secluded haunts, but he seems equally at home in maples and elms about our houses, or even in the more quiet village streets. He is therefore more often heard than his mysterious relative, and, as a voice, is familiar to many who do not know the singer's name.

The call-notes of the Wood Thrush are a liquid *quirt* and sharp *pit-pit*. The latter is an alarm note, which, when the bird fears for the safety of its young, is uttered with much increased force and rapidity. It can be closely imitated by striking two large pebbles together.

The Alarm

PLATE LXXIV.

WOOD THRUSH.

Length, 8·30 inches. Upper parts bright, rusty cinnamon, brightest on back and crown ; under parts white everywhere, except center of belly, with large, rounded black spots.

18

The song of the Wood Thrush is wholly unlike that of the Veery. It opens with the flutelike notes, and is sung disconnectedly, being broken by pauses and by low notes, audible only when one is near the singer.

Come to me,

The Hermit Thrush is a more northern bird than either the Veery or the Wood Thrush. It rarely nests at sea level south of Vermont or northern Michigan, but in the higher portions of Massachusetts and on the crests of the Catskills and Alleghanies in Pennsylvania, it is also found breeding. It winters from southern Illinois and New Jersey southward to the Gulf, it being the only member of its genus to inhabit the eastern United States at that season. Its spring migrations occur between April 5 and May 10, and in the fall we see it from October 15 to November 25, while occasionally it may winter.

Hermit Thrush,
Turdus aonalaschkæ
pallasii.
Plate LXXV.

During its migrations the Hermit Thrush usually frequents woodlands, where it may often be seen on or near the ground. Like the Veery, it is a ground-nester, and its eggs, though slightly lighter in color, resemble those of the Veery and Wood Thrush in being plain, bluish green. When alighting, the Hermit has a characteristic habit of gently raising and lowering its tail, and at the same time uttering a low *chuck*. Sometimes it sings during the winter, in Florida, and also while migrating: but if you would hear this inspired songster at his best, you must visit him in his summer home. The Hermit's song resembles that of the Wood Thrush in form, but it is more tender and serene. O spheral, spheral! O holy, holy! Mr. Burroughs writes the its opening notes, and there is something about the words which seems to express the spirit of heavenly peace with which the bird's song is imbued.

PLATE LXXV.

HERMIT THRUSH.

Length, 7·15 inches. Upper parts and wings dark cinnamon-brown; tail bright reddish brown; under parts white; breast spotted with black; sides washed with brownish; belly white.

259

It seems a long step from these gentle, refined Thrushes to their comparatively prosaic cousin, the famil-

Robin,
Merula migratoria.

iar Robin. But the Robin has his place, and in March his cheery song is quite as effective as the Hermit's hymn in June.

During the summer Robins are distributed through-out North America from the Gulf States and southern end of the Mexican tableland, northward to Labrador and Alaska. In the winter they may be found in numbers from Virginia southward, small flocks and single birds being occasionally met with as far north as Massachu-setts. Robins are among our earliest migrants, appear-ing in the vicinity of New York city between February 20 and March 1. Nesting is begun about April 15, the mud-lined nest and greenish blue eggs being too well known to require description. Two, or even three broods may be raised. In June, the young of the first brood with some adult males resort each night to a chosen roost, often frequented by many thousands of birds.

The fall migration begins in September, but the birds are with us in roving bands until December.

About the time that we first hear the Robin's ringing welcome to spring we may listen for the Bluebird's more

Bluebird,
Sialia sialis.

gentle greeting. Doubtless the bird has been with us all winter, for Blue-birds winter in small numbers as far north as southern Connecticut, often living near groves of cedars, which offer them both food and shelter. In the Southern States they are far more abundant at this season, gathering in flocks containing hundreds of indi-viduals.

The Bluebird is the first of our smaller birds to begin housekeeping, and early in April it may be seen pro-specting about the site of last year's nest in a bird box or hollow tree, and the bluish white eggs will probably be laid before the middle of the month.

No bird's song is more associated with the return of spring than the Bluebird's; nor is there a bird's note more expressive of the passing season than the Bluebird's autumn call of *far-away, far-away.*

INDEX.

APPENDIX.

PREFACE TO TEACHERS' EDITION.

WHILE the time available for zoölogical studies in our schools is too limited to permit of more than the treatment in outline of most of the classes of animals, the fact is now recognized that birds possess unusual claims to our attention. They are practically the only ones of the higher animals with which we may come in contact daily. Our large mammals have either been exterminated or driven from the vicinity of our homes, while most of the smaller species are nocturnal, and, therefore, rarely seen. Reptiles and batrachians are difficult to observe and are not popular; while fishes, from the nature of their haunts, can be studied only under certain conditions. Birds, however, are everywhere: in field and wood and sky, in our orchards and gardens; and some of them are with us at all seasons.

But birds' merits do not consist merely in their abundance. In beauty of plumage, grace of motion, and vocal ability they are without rivals; in their migration, mating, and nesting habits they not only display unusual intelligence, but exhibit human traits of character that create within us a feeling of kinship with them, and thus increase our interest in and love for them. Furthermore, as with increasing knowledge we begin to realize their economic value, we are more than ever im-

pressed with the importance of becoming acquainted
with them.

Still, it will be obviously impossible for the student to cover the whole field of ornithology, and
the question arises, to what phase of the subject
he should give special attention.

There are teachers who believe that classification
is the principal object of natural history study, and
the aim and end of their instruction is to teach the
pupil the names of Orders and Families, and the
characters on which they are based. So far as birds
are concerned, the plan is excellent as a preliminary
step, but to my mind it is of infinitely greater importance to be able to recognize a Wood Thrush or a
Veery than to define the Lamellirostral Grallatores.

In this book structure and classification have,
therefore, been subordinated to matter which will
be of practical assistance to the student in identifying the birds about his home, and in teaching him
to appreciate their economic, æsthetic, and scientific
value.

If he lives in the country, this information may be
of service to him daily ; and this, it seems to me, is
a far more profitable kind of ornithology than that
which treats only of " Orders," and " Families,"
and " leading types " which he will probably never
see outside of a museum or a zoölogical garden.

Acting on this belief, I have written of the living,
rather than of the dead bird, and no attempt, therefore, has been made to describe the anatomy of
birds, but, in preference, the questions of economics,
æsthetics, form and habit, color, migration, song,
nesting, etc., have been dwelt on with the ob-

ject of both cultivating and directing the student's
powers of observation. In order, however, to give
him some idea of the bird's place in Nature, the subjects of relationships and classification have been
touched on briefly. Then follow a series of objec-
tive, seasonal lessons which are the main feature of
the book. The advantages of studying birds under
seasonal groupings are two-fold. First, by elimi-
nating species which are absent, it greatly simplifies
the question of identification. Second, it is more
real. If the student can be told that a certain species will doubtless arrive from the south the same
day on which he is reading about it, his interest in
the subject will be at once increased : it becomes a
matter of contemporary history. Furthermore, by
studying the birds with the seasons, we learn in the
beginning to properly associate them with certain
accompanying natural phenomena, and their comings and goings become significant events in our
calendar.

As we become familiar with birds, and learn to
recognize them, the question of identity will no
longer remain a bar to our better acquaintance, and
our interest in them will deepen. We shall begin
to inquire into the questions of form and habit,
color, migration, song, nesting, etc.; and as a guide
to observations of this character, there are given a
series of lessons treating of the philosophic or subjective side of bird-study, the wide scope of which
will be readily appreciated.

F. M. C.

AMERICAN MUSEUM OF NATURAL HISTORY,
November, 1898.

QUESTIONS ON CHAPTER I.

THE BIRD : ITS PLACE IN NATURE AND RELATION TO MAN.

The Bird's Place in Nature (see Chapter I, pages 1–5).—
How many species of birds are known ? In what class are
they placed ?* Name the classes of higher animals; that is,
mammals, fishes, and reptiles. In what respect does the
class birds differ from all the other higher classes of animals ?† What place does the class birds occupy in the scale
of life ? To what class are they most nearly related ?

Are birds the only higher animals that fly ? Are they the
only ones that lay eggs ? The only ones that incubate ?
What is the temperature of birds ? Of mammals ? Of
reptiles ? Have any living birds teeth ? What is the chief
peculiarity of birds ? From what kind of ancestors are
birds believed to have descended ? On what evidence is
this belief based ?

Describe the Archaeopteryx. Where was it found ? In
what geologic age did it live ? Do birds vary much in
structure ? In habit ?

Mention some varying habits of birds.

Economic Relations of Birds to Man (see Chapter I,
pages 5–9).—In what ways are birds useful to man ? What
loss are insects estimated to inflict on our agricultural interests annually ? What birds catch insects on the wing ?
In the foliage ? On the tree trunks ? What kinds feed on
terrestrial insects ? Describe Mr. Forbush's observations

---

* The teacher should define the meaning of "Class" : as, for example, the class Mammalia, the class Reptilia, etc.

† For example, such extreme representatives of the class Aves
as the Hummingbird and Ostrich, resemble each other in more
respects than do, for instance, the Bat and the Elephant in the
class Mammalia.

on the food of the Chickadee. What was found in the stomach of a Yellow-billed Cuckoo? Of a Robin? Are most Hawks and Owls beneficial birds? What forms the largest part of the food of the Red-shouldered Hawk? What was found in the castings* of the Barn Owl? What State offered a reward for Hawks and Owls? What loss is estimated to have resulted? Why are seed-eating birds of economic value? What birds are useful as scavengers? What was the result of killing birds on the Yucatan Coast?†

Æsthetic Relations of Birds to Man (see Chapter I, pages 10–13).—After learning to know birds, what æsthetic characters shall we find that they possess? Mention several birds of beautiful plumage. Several of unusually graceful flight. Several musical birds. What human traits of character are exhibited by birds? What pleasure is to be derived from acquaintance with birds? Is their study restricted to any special season? In what manner will birds appeal to us most strongly?

Does familiarity with their notes increase the pleasure we receive from birds? Is this the result of association? In what manner?

---

* Undigested pellets of hair, feathers, and bones, which are ejected at the mouth by Owls and some other birds.

† The lesson may be extended by exhibiting plates of, and describing the food of, some of our most useful birds, such as the Swallows (Plate I), Herring.Gull (Plate IV), Turkey Vulture (Plate LXXVIII), Red-shouldered Hawk (Plate XIV), Marsh Hawk (Plate XV), Sparrow Hawk (Plate XVI), Short-eared Owl (Plate XIX), Screech Owl (Plate XX), Barred Owl (Plate XXI), Yellow-billed Cuckoo (Plate XXII), Downy Woodpecker (Plate XXIV), Flicker (Plate XXVI), Nighthawk and Whip-poor-will (Plate XXVII), Kingbird (Plate XXX), Cedar Waxwing (Plate LVII), Vireos (Plate LIX), Chickadee and Brown Creeper (Plate LXX), and Nuthatches (Plate LXXI).

Identification (see Chapter VII).—As a preliminary step to exercises in identification the student
should learn to name the parts of a bird's plumage
as they are given in Figure 25. The teacher should
then select a plate of a land-bird, and placing it at
a distance of from twenty to thirty feet from the
pupil, have him write a one-minute description of it.
This description should include the bird's approximate length,* color of crown, back, tail, wings,
throat, breast, and abdomen. It is well to have
a blank prepared and ready to fill in with the
descriptions of the parts named. To this may
be added any particular characters of form (*e. g.*,
crests, long tail, etc.) or color (*e. g.*, face or rump
marks, etc.).

With this description in hand the student should
then turn to the key on page 76. This is primarily
designed to identify birds in Nature, and its major
divisions are based on the most striking habits of the
birds. This, however, would not be appreciable in
the bird plate, and the teacher should, therefore,
designate in which of the three principal groups the
bird belongs. The pupil should then proceed with

---

* A Robin is ten inches, an English or House Sparrow six and
one-quarter inches, in length. Mental comparison with either of
these familiar birds will enable one to readily estimate the length
of any of our Passeres.

the identification of the bird, as explained on
page 75.

Exercises of this nature should be repeated until
the student can describe birds quickly and accurately
and has thoroughly mastered the use of the key.

If possible, this class-room work should be frequently supplemented by observations in the field.
When the country is not available, large parks often
prove by no means poor substitutes, and during the
migrations they are frequently thronged with birds.

Even when field lessons are out of the question, it
is strongly advised that the studies of certain birds
be made during the season when they are present.
The best plan is to begin in December with the birds
which are with us throughout the year, or the Permanent Residents, adding the Winter Visitants in
January and February. As the migrants from the
south appear, they may form the subjects of the
month's lessons, and the course ends naturally in
June, when all the summer birds have arrived.

This method associates the birds with their respec-
tive seasons, and for the field student is particularly
advantageous. He takes up the subject at a time
when the comparatively small number of birds present greatly simplifies the question of identification,
and before the first migrants arrive in March, should
have become familiar with the commoner Permanent
Residents and Winter Visitants.

When field work is practicable, each student should
keep a record of the birds observed. Notes of this
kind, made during the migration, are particularly
interesting. They may be entered on a large page
ruled in squares, similar in style to those of a roll-call

book. The bird's name is entered at the left side of the page, the date at the top, and the record for the day is placed in the square opposite the bird's name and below the date. It may consist simply of a check or mark indicating that the species was seen, but preferably should give the approximate number of individuals observed ; whether the species was heard singing; whether observed in flocks; and any other information which can be easily and intelligibly abbreviated.

A journal should be kept in which to write a more detailed account of the day's experiences. These may also form the subject of compositions, and the class-room work should now include comparison and discussion of observations made in the field. Compositions may also be written on certain species, when the outline of a bird's biography, given on page 73, will furnish suggestions as to the heads under which the subject may be treated.

Later, the philosophic or subjective side of bird-study may be considered, and compositions written on structure and habit, color, migration, nesting, etc.

As a definite guide to seasonal bird-studies in the middle Eastern States, the following outline of the bird-life of a year is given. It is based on observations made in the vicinity of New York city, and includes all the land-birds and the commoner water-birds inhabiting this region. It may be prefaced by a definition of the four groups in which our birds are placed, according to the manner of their occurrence (see page 53), as follows:

PERMANENT RESIDENTS.

Permanent Resident species are those which are represented in the same locality throughout the year. This does not imply that the same individuals live in one locality continuously; few of our birds being Permanent Residents in the strict sense of the word. Doubtless, Ruffed Grouse, Bob-whites, and possibly a few other species pass their lives in the vicinity of their birth; but most species ranked as Permanent Residents are, in fact, more or less migratory. Thus, in the vicinity of New York city, Chickadees and Bluebirds are found every month of the year; but in October, many migrants of both species may be seen, and it is probable that we then receive our Winter Residents of these species, while the birds that were with us during the summer go farther south to pass the winter.

List of Permanent Residents.

Plate No.		Plate No.	
77.	Bob-white.	19.	Short-eared Owl.
12.	Ruffed Grouse.	21.	Barred Owl.
14.	Red-shouldered Hawk.		Great Horned Owl.*
	Red-tailed Hawk.	24.	Downy Woodpecker.
	Broad-winged Hawk.*		Hairy Woodpecker.
15.	Marsh Hawk.	25.	Red-headed Woodpecker.
16.	Sparrow Hawk.		(Irregular.)
	Duck Hawk.*	26.	Flicker.
17.	Sharp-shinned Hawk.		Prairie Horned Lark.*
	Cooper's Hawk.*	80.	American Crow.
79.	Bald Eagle.*		Fish Crow.
20.	Screech Owl.	81.	Blue Jay.
	Long-eared Owl.*		Starling. (Introduced.)

* Not common.

Plate
No.
41. Song Sparrow.
   House or English Sparrow. (Introduced.)
52. American Goldfinch.
   European Goldfinch. *(Introduced.)
53. Purple Finch.
84. Cardinal * (From New York city and southward.)

Plate
No.
57. Cedar Waxwing.
   Carolina Wren.* (From New York city and southward.)
70. Chickadee.
   Tufted Titmouse.* (From New York city and southward.)
71. White-breasted Nuthatch.
90. Bluebird.

WINTER VISITANTS.

The term Winter Visitant, like that of Summer Resident, is not used in an exact sense, but is applied to birds that arrive from the north in the fall, pass the winter with us, and return to their more northern homes the following spring. Most of them arrive late in September and depart in April.

In addition to these regular Winter Visitants, there sometimes occur irregular Winter Visitants, whose coming cannot be foretold. Absent some winters, they may be abundant others ; their presence or absence being apparently governed by the supply of food to the northward. When this fails, they sweep southward in enormous numbers, becoming common in localities where they are usually rare or unknown. Pine Grosbeaks, Crossbills, and Redpolls are irregular Winter Visitants.

LIST OF WINTER VISITANTS.

Plate
No.
4. Herring Gull.
   Saw-whet Owl.*
34. Horned Lark.

Plate
No.
48. Junco.
49. Tree Sparrow.
   Pine Siskin.†

* Not common.   † Irregular.

Plate
No.

50. Redpoll.†
50. Snowflake.†
    Lapland Longspur.*
51. American Crossbill.†
    White-winged Crossbill.*
51. Pine Grosbeak.†

Plate
No.

58. Northern Shrike.*
    Winter Wren.
70. Brown Creeper.
71. Red-breasted Nuthatch.†
72. Golden-crowned Kinglet.

TRANSIENT VISITANTS.

This group includes species which pass us each spring in going to their more northern nesting grounds, and which visit us again in the fall in returning to their more southern winter homes.

The earlier Transient Visitants—for example, Wilson's Snipe and the Fox Sparrow—may remain with us a month or six weeks should the season be backward, but the later arrivals—for instance, the Warblers of May, who arrive when the weather is comparatively settled—pass us in a week or ten days.

Most of our Transient Visitants are Ducks, Geese, Snipe, and Plover, who travel far northward beyond the haunt of man to breed in security; and Warblers and Thrushes, who nest in the great spruce and balsam forests of northern New England and Canada.

LIST OF TRANSIENT VISITANTS.

Plate
No.

2. Pied-billed Grebe.
3. Loon.
5. Blue-winged Teal.
5. Green-winged Teal.
5. Pintail.
5. Canada Goose.

Plate
No.

8. American Coot. *
9. Wilson's Snipe.
10. Semipalmated Sandpiper.
    Solitary Sandpiper.
10. Semipalmated Plover.
19. Short-eared Owl.

---

* Not common.   † Irregular.

Plate
No.

19. Yellow-bellied Woodpecker.
Olive-sided Flycatcher.*
Yellow-bellied Flycatcher.
Traill's Flycatcher.*
Rusty Blackbird.
Bronzed Grackle.
Nelson Sharp-tailed Sparrow.*
Acadian Sharp-tailed Sparrow.*
White-crowned Sparrow.*
Lincoln's Sparrow.*
47. Fox Sparrow.
Philadelphia Vireo.*
Blue-headed Vireo.
Nashville Warbler.
Tennessee Warbler.
Cape May Warbler.*

Plate
No.

Black-throated Blue Warbler.
61. Myrtle Warbler.
Magnolia Warbler.
Bay-breasted Warbler.*
Black-poll Warbler.
Blackburnian Warbler.
61. Black-throated Green Warbler.
Yellow Palm Warbler.
Small-billed Water Thrush.
Connecticut Warbler.*
Mourning Warbler.*
Wilson's Warbler.
Canadian Warbler.
Titlark.
72. Ruby-crowned Kinglet.
Gray-cheeked Thrush.*
Bicknell's Thrush.*
Swainson's Thrush.
75. Hermit Thrush.

SUMMER RESIDENTS.

The term Summer Resident is applied to those species which come to us from the south in the spring, rear their young, and return to the south in the fall. Summer Residents, therefore, are present not only during the summer months, but may arrive in late February or early March, and remain until late November or early December.

As a rule, the first species to come in the spring are the last to leave in the fall, while the later arrivals are among the first departures.

Species that come in March or early April are

* Not common.

present a month or more before beginning to nest, but those that come in May may be found nest-building within a few days after their arrival.

LIST OF SUMMER RESIDENTS.

Plate No.		Plate No.	
	Laughing Gull.*	35.	Baltimore Oriole.
1 .	Common Tern.*	36.	Orchard Oriole.
.	Wood Duck.*	82.	Red-winged Blackbird.
.	Great Blue Heron.*	37.	Purple Grackle.
.	Little Green Heron.	38.	Bobolink.
6.	Black-crowned Night Heron.	39.	Meadowlark.
		40.	Cowbird.
7.	American Bittern.*		Grasshopper Sparrow.
	Least Bittern.		Henslow's Sparrow.*
8.	Clapper Rail.	42.	Swamp Sparrow.
	King Rail.*	43.	Field Sparrow.
	Virginia Rail.*	44.	Vesper Sparrow.
76.	Woodcock.	45.	Chipping Sparrow.
11.	Spotted Sandpiper.	54.	Rose-breasted Grosbeak.
11.	Killdeer.*	55.	Towhee.
13.	Mourning Dove.	83.	Indigo Bunting.
18.	Osprey.	85.	Scarlet Tanager.
	Barn Owl.*	1.	Barn Swallow.
22.	Yellow-billed Cuckoo.		Rough-winged Swallow.*
	Black-billed Cuckoo.	1.	Cliff Swallow.
23.	Belted Kingfisher.	1.	Bank Swallow.
27.	Nighthawk.	1.	Tree Swallow.
27.	Whip-poor-will.	86.	Purple Martin.
28.	Chimney Swift.	59.	Red-eyed Vireo.
29.	Ruby-throated Humming-bird.		Warbling Vireo.
		59.	Yellow-throated Vireo.
30.	Kingbird.		White-eyed Vireo.
31.	Crested Flycatcher.	60.	Black and White Warbler.
32.	Phœbe.		Blue-winged Warbler.
	Least Flycatcher.		Worm-eating Warbler.*
	Acadian Flycatcher.	87.	Yellow Warbler.
33.	Wood Pewee.		

*Not common.

Plate
No.

Chestnut-sided Warbler.
Prairie Warbler.
62. Redstart.
Hooded Warbler.
63. Oven-bird.
Louisiana Water Thrush.
64. Maryland Yellowthroat.
Kentucky Warbler.*
65. Yellow-breasted Chat.

Plate
No.

88. Catbird.
67. Brown Thrasher.
68. House Wren.
Short-billed Marsh Wren.
69. Long-billed Marsh Wren.
73. Veery.
74. Wood Thrush.
89. Robin.
90. Bluebird.

* Not common.

JANUARY.

PROBABLY during no other month is there less movement among our birds than in January. All the regular Winter Visitants have come ; the Fall Migrants, which may have lingered until December, have gone, and the earliest Spring Migrants will not arrive before the latter part of February or early in March. In fact, January is the only month in the year in which, as a rule, some birds do not arrive or depart. This rule, however, may be broken by such irregular birds as the Pine Grosbeak and Redpoll, and, south of the latitude of New York city, by the Snowflake and Crossbill, birds which are wholly absent some winters and abundant others.

The only birds usually to be found in January, therefore, are the Permanent Residents and regular Winter Visitants. Singing, mating, nesting, molting, migrating—events which, in their season, play so important a part in a bird's life—do not concern the birds of January. With them food is the one important question, and their movements at this season are governed solely by the food supply. Snow may fall and winds may blow, but as long as the birds find sufficient to eat, they give small heed to the weather. Where seed-bearing weeds are accessible, there we may look for Juncos and Tree Sparrows ; a cedar-tree filled with berries often tempts Robins, Bluebirds, and Waxwings to winter near it. I recall a sheltered pile of buckwheat chaff at Englewood, N. J., which furnished food for a small flock of Mourning Doves all one winter. In Central Park, New York city, a Mockingbird, who had evidently escaped from a cage, fed upon the berries of a privet tree, and survived in apparent comfort the most severe winter weather. Food, therefore, rather than temperature, is the all important factor in a bird's life at this season.

BIRDS OF THE MONTH.

PERMANENT RESIDENTS (see page 6).

WINTER VISITANTS (see page 7).

FEBRUARY.

The conditions prevailing in the bird world during January will be practically unchanged until the latter part of February. Then, should there be a period of milder weather, we may expect to hear the Song Sparrow and Bluebird inaugurate the season of song. An unusually warm day, earlier in the month, may have tempted either or both of these birds to prematurely welcome spring, but as a rule we do not hear them until late in February, and then only under favorable conditions.

The song of these birds bids us keep watch for the earliest migrants, the Robin, Purple Grackle, and

Red-winged Blackbird, birds which pass the winter such a short distance south of us that they appear at the first sign of returning spring.

Further confidence in the growth of the new year is shown by the Great Horned Owl, one of our less common species, who begins nesting late in February or early in March.

But in spite of these movements among the birds, February is, generally speaking, a winter month, and it is only in exceptional years that we shall find much change in our avifauna.

BIRDS OF THE MONTH.

PERMANENT RESIDENTS (see page 6).

WINTER VISITANTS (see page 7).

MIGRANTS.

February 15 to 28, in favorable seasons.

Plate No.	Plate No.
37. Purple Grackle.	82. Red-winged Blackbird.
Rusty Blackbird.*	89. Robin.

BIRDS NESTING.

Great Horned Owl—February 20-28.

MARCH.

While March is sure to witness a general northward movement among the birds, the date of their arrival is as uncertain as the weather of the month itself. Continued severe weather prevents their advance, which a higher temperature as surely occasions. It is well, therefore, to watch closely the weather predictions, knowing that birds will quickly

---

* Transient Visitant passing further north.

follow in the wake of a warm wave. When the ice leaves our bays, ponds, and rivers, Ducks and Geese will appear. Even before this event, the Grackles, Red-winged Blackbirds, and Robins will come in flocks and in song, and singing will become general with the Song Sparrows and Bluebirds, whose numbers will be greatly increased. When successive thaws have rendered the earth soft enough for the Woodcock's probe, we may expect to find him in favorable localities, searching for his fare of earthworms. With the advent of insects, we may look for their enemy, the Phœbe, and when the frogs begin peeping in the ponds and marshes, we shall know that the spring migration is well under way, and that Meadowlarks, Cowbirds, and other March Migrants may be found for the seeking.

To the lover of bird music the event of the month will be the first Fox Sparrow song; heard at this season it is a thrilling performance.

The weather which hastens the arrival of birds from the south, also prompts certain of our Winter Visitants to begin their northward journey, and after March we do not often see Redpolls, Snowflakes, and Northern Shrikes.

BIRDS OF THE MONTH.

PERMANENT RESIDENTS (see page 6).

WINTER VISITANTS (see page 7).

The following will leave for the north:

Plate No.		Plate No.	
34.	Horned Lark.	51.	Pine Grosbeak.
50.	Redpoll.	58.	Northern Shrike.
50.	Snowflake.		

MIGRANTS.

Plate No.

Appearing when the Ice leaves the Bays and Rivers.

1. Loon.*
5. Pintail.*
5. Mallard.*
5. Green-winged Teal.*
5. Blue-winged Teal.*
5. Canada Goose.*

March 1 to 10.

37. Purple Grackle.
82. Red-winged Blackbird.
     Rusty Blackbird.*
89. Robin.

Plate No.

March 10 to 20.

76. Woodcock.
32. Phœbe.
39. Meadowlark.
40. Cowbird.
47. Fox Sparrow.*

March 20 to 31.

9. Wilson's Snipe.*
23. Kingfisher.
13. Mourning Dove.
42. Swamp Sparrow.
46. White-throated Sparrow.*

BIRDS NESTING.

March 1 to 15.

21. Barred Owl.

March 15 to 31.

     Duck Hawk.
     Carolina Wren.

APRIL.

In early April, the developments in the vegetable world, which the most casual observer cannot fail to see, are accompanied by corresponding, but less noticed, activities in the world of birds. The appearance of the skunk cabbage, the blossoming of the pussywillow and early wild flowers soon become common knowledge; but the arrival of the Vesper, Field, and Chipping Sparrows; of Tree Swallows, Myrtle Warblers, and Hermit Thrushes, is known to comparatively few. Still, to the bird-lover, the return of these feathered friends is of even greater interest than the blooming of trees and plants.

---

* Transient Visitant passing further north.

The migratory movement rapidly grows in strength, and, during the latter part of the month, one may expect to see new comers almost daily.

It will be noted that the earlier migrants of the month are all seed-eaters, who return just in time to help the remaining Winter Visitants harvest what is left of the preceding year's crop of seeds. Later, certain insectivorous birds which catch their prey on the wing are found; for example, the Swallows, Swift, and Nighthawk.

BIRDS OF THE MONTH.

PERMANENT RESIDENTS (see page 6).

REMAINING WINTER VISITANTS (see page 7).

The following will leave for the north:

Plate No.	Plate No.
48. Junco.	70. Brown Creeper.
49. Tree Sparrow.	71. Red-breasted Nuthatch.
Winter Wren.	72. Golden-crowned Kinglet.

MIGRANTS.

April 1 to 10.
2. Pied-billed Grebe.
6. Great Blue Heron.*
6. Black-crowned Night Heron.
18. Osprey.
44. Vesper Sparrow.
Savanna Sparrow.
43. Field Sparrow.
45. Chipping Sparrow.
1. Tree Swallow.
61. Myrtle Warbler.*
American Pipit.*

75. Hermit Thrush.*

April 10 to 20.
7. American Bittern.
6. Green Heron.
8. Clapper Rail.
Yellow-bellied Woodpecker.*
1. Barn Swallow.
Yellow Palm Warbler.*
Pine Warbler.
Louisiana Water Thrush.*
72. Ruby-crowned Kinglet.*

---

* Transient Visitant passing further north.

Plate No.

April 20 to 30.

11. Spotted Sandpiper.
10. Semipalmated Sandpiper.*
27. Whip-poor-will.
28. Chimney Swift.
    Least Flycatcher.
55. Towhee.
    Blue-headed Vireo.

Plate No.

86. Purple Martin.
1. Cliff Swallow.
1. Bank Swallow.
    Rough-winged Swallow.
60. Black and White Warbler.
61. Black-throated   Green
    Warbler.
67. Brown Thrasher.

BIRDS NESTING.

(In addition to the species which began to nest in March, all of which will have eggs or young in April, the following may be found nesting) :

Plate No.

April 1 to 15.

76. Woodcock.
14. Red-shouldered Hawk.
20. Screech Owl.
    Red-tailed Hawk.
80. American Crow.
    Long-eared Owl.
90. Bluebird.

Plate No.

April 15 to 30.

71. White-breasted Nut-
    hatch.
89. Robin.
13. Mourning Dove.
37, 91. Purple Grackle.
32. Phœbe.
41. Song Sparrow.

MAY.

As the season advances, marked changes in temperature are less likely to occur, and the migration becomes more regular and continuous. In February and March there may be two weeks or more variation in the times of arrival of the same species in different years ; in May we can expect to find a given species within a day or two of a certain date. Nevertheless, we shall find the force of the migratory current still closely dependent on meteorologic

---

* Transient Visitant passing further north.

conditions, and under the encouragement of a high temperature we may be visited by bird "waves" which flood the woods with migrants. Birds are then, doubtless, more abundant than at any other period of the year. As many as ten species may be noted as arriving on the same day, while the number of individuals observed may almost exceed calculation. At this season it is not unusual to observe from sixty to eighty species of birds during a few hours' outing, and Mr. W. L. Dawson records that, with Prof. Lynds Jones of Oberlin College, he recorded twelve species of water birds and ninety species of land birds in one day of field work in Lorain County, Ohio.

After the fifteenth of the month, birds begin to decrease in number, the Transient Visitants passing further north, and by June 5 our bird-life is composed of Permanent Residents and Summer Residents.

It will be noticed that with few exceptions the birds arriving in May are insectivorous ; particularly those insect-eating birds which obtain their food from the vegetation. Thus, no sooner are the unfolding leaves and opening blossoms exposed to the attack of insects than the Warblers and Vireos appear to protect them, and the abundance of these small birds is the distinctive feature of the bird-life of the month.

Their diminutive size, activity, and the persistence with which they remain in the tree-tops render their identification in life by no means an easy matter, and more than any of the other land birds they test the patience of the field student.

May is preeminently the month of courtship, which
finds expression chiefly in song.   Many species begin
to nest in May, but the nesting season reaches its
height the following month.

BIRDS OF THE MONTH.

PERMANENT RESIDENTS (see page 6).

MIGRANTS.

*Plate
No.*

May 1 to 10.

10. Common Tern.
  Solitary Sandpiper.*
10. Semipalmated Plover.*
22. Yellow-billed Cuckoo.
  Black-billed Cuckoo.
27, 91. Nighthawk.
29, 91. Ruby-throated Hummingbird.
31, 91. Crested Flycatcher.
30. Kingbird.
35. Baltimore Oriole.
36. Orchard Oriole.
38, 91. Bobolink.
  Grasshopper Sparrow.
83. Indigo Bunting.
54. Rose-breasted Grosbeak.
85. Scarlet Tanager.
59, 91. Red-eyed Vireo.
  Warbling Vireo.
59. Yellow-throated Vireo.
  White-eyed Vireo.
  Nashville Warbler.*
  Blue-winged Warbler.
  Parula Warbler.
  Black-throated Blue Warbler.*

*Plate
No.*

  Magnolia Warbler.*
  Chestnut-sided Warbler.
  Prairie Warbler.
  Small-billed Water
  Thrush.*
  Hooded Warbler.
87. Yellow Warbler.
64, 98. Maryland Yellowthroat.
65. Yellow-breasted Chat.
63, 91, 97. Oven-bird.
62. Redstart.
68, 91. House Wren.
88, 99. Catbird.
74, 100. Wood Thrush
73. Veery.

May 10 to 20.

33. Wood Pewee.
  Acadian Flycatcher.
  Yellow-bellied Flycatcher *
  White-crowned Sparrow.*
  Golden-winged Warbler.*
  Tennessee Warbler.*
  Worm-eating Warbler.
  Cape May Warbler.*
  Blackburnian Warbler.*
  Bay-breasted Warbler.*

* Transient Visitant passing further north.

Plate
No.

Black-poll Warbler.*
Wilson's Warbler.*
Canadian Warbler.*
69. Long-billed Marsh Wren.
Short-billed Marsh Wren.

Plate
No.

Olive-backed Thrush.*
Gray-cheeked Thrush.*
Traill's Flycatcher.*
Mourning Warbler.*
Bicknell's Thrush.*

BIRDS NESTING.

(In addition to the species which began to nest in May, all of which will have eggs or young in June, the following may be found nesting:)

Plate
No.

May 1 to 10.
5. Wood Duck.
6. Green Heron.
6. Black-Crowned Night Heron.
8. Clapper Rail.
11, 92. Killdeer.
12. Ruffed Grouse.
Cooper's Hawk.
16. Sparrow Hawk.
18. Osprey.
23. Kingfisher.
26. Flicker]
44. Vesper Sparrow.
Savanna Sparrow.
84. Cardinal.
1. Barn Swallow.
King Rail.

May 10 to 20.
Virginia Rail.
25. Red-headed Woodpecker.
29, 91. Ruby-throated Hummingbird.
Acadian Flycatcher.
81. Blue Jay.

Plate
No.

Fish Crow.
82. Red-winged Blackbird.
39. Meadowlark.
45. Chipping Sparrow.
43. Field Sparrow.
42. Swamp Sparrow.
55. Towhee.
44. Rose-breasted Grosbeak.
1. Tree Swallow.
1. Bank Swallow.
Blue-winged Warbler.
Hooded Warbler.
60. Black and White Warbler
62. Redstart.
Worm-eating Warbler.
63, 91, 97. Oven-bird.
67. Brown Thrasher.
88, 99. Catbird.
70. Chickadee.
74, 100. Wood Thrush.
73. Veery.

May 20 to 31.
Least Bittern.
11. Spotted Sandpiper.

* Transient Visitant passing further north.

JUNE.

After June 5 we may be reasonably sure that every bird seen has, or has had, a nest in our vicinity. Several of the birds which began nesting in April—for instance, the Phœbe, Song Sparrow, Robin, and Bluebird—will rear second broods in June, while the young of other April nesting birds, such as the Red-shouldered Hawk, Screech Owl, and Crow, may not leave the nest until June. All the birds that began nesting in May will still be occupied with household affairs in June; and when we add to these the late-breeding species which wait for June before settling their domestic arrangements, it will be seen that among birds June is the home month of the year.

Nest-building, egg-laying, incubating, and the care of the young now make constant and exceptional demands on birds, who, in response, exhibit traits which at other times of the year they give no evidence of possessing. Singing now reaches its highest development, and certain call-notes are heard only at this season. The numberless actions incident to courtship; the intelligence displayed in nest-building; the choice of special food for the young ; the devotion which prompts the parents to recklessly expose themselves in the protection of their offspring—all these manifestations of the bird-mind may be observed in June.

A feature of the bird-life of the month is the formation, usually in young second-growth woods, of roosts which are nightly frequented by the now fully grown young of such early-breeding birds as the Purple Grackle and Robin. When a second brood is raised, as with the Robin, the young of the first brood may be accompanied to the roost by the male parent, but in the one-brooded Grackle the roost is used by both adults and young.

BIRDS OF THE MONTH.

PERMANENT RESIDENTS (see page 6).

SUMMER RESIDENTS (see page 10).

BIRDS NESTING.

(In addition to the species which began to nest in May, all of which will have eggs or young in June, the following may be found nesting:)

Plate No.		Plate No.	
	June 1 to 10.		85. Scarlet Tanager.
	Laughing Gull.		
27, 91.	Nighthawk.		*June 10 to 20.*
27.	Whip-poor-will.	10.	Common Tern.
31, 91.	Crested Flycatcher.	57, 91.	Cedar Waxwing.
33, 91.	Wood Pewee.	52.	American Goldfinch.

July.

The full development of the bird year is attained in June, and as early as the first week in July, when, among some migratory birds, there are evidences of preparation for the journey southward, the season begins to wane. The young of certain species which rear but one brood have now left the nest, and, accompanied by their parents, wander about the country. In localities which we had thoroughly explored in June, we may therefore find species not met with before. In some cases, these families join others of their kind, forming small flocks, the nucleus of the great gatherings seen later. Examples are Grackles, Red-winged Blackbirds, and Tree Swallows. The latter rapidly increase in number, and by July 10 we may see them, late each afternoon, flying to their roosts in the marshes.

During the first week in the month we shall also find that certain birds have concluded their season of song.

Bobolinks and Red-winged Blackbirds are rarely heard after the tenth of the month; their young are reared, the cares of nesting-time have passed, and, with other one-brooded birds, they begin to renew their worn breeding plumages by molting. After the fifteenth we miss the voices of the Veery, Orchard and Baltimore Orioles, Chat, Brown Thrasher, and others. But in place of the songs of these more prominent members of the bird choir, we notice the calls of certain young birds who, long after they have left the nest, are still dependent on their parents; thus the squawkings of young Crows and trem-

ulous cries of immature Baltimore Orioles are characteristic of the season.

BIRDS OF THE MONTH.

Permanent Residents (see page 6).

Summer Residents (see page 10).

August.

With the majority of our nesting birds, family
cares are ended in August, and at this season they
completely renew their worn plumages by molting.
As every keeper of cage-birds well knows, this is a
trying period in a bird's life. Wild birds molt more
quickly than caged ones, and it is possible that the
physical strain to which the growth of new feathers
subjects them may be more severe. However this
may be, birds when molting are less in evidence than
at any other time. What becomes of many of our
birds in August, it is difficult to say. Baltimore
Orioles, for example, are rarely seen from August 1
to 20, but after the latter date they reappear clad
in new plumage and are then in nearly full song.
So apparently complete is the disappearance of
birds in August that before the fall migration daily
brings new arrivals from the north, one may spend
hours in the woods, and hear only the Red-eyed
Vireo and Wood Pewee, August's own songsters.
Late in the month, migrants from the north will be
found travelling through the woods in small companies, but the characteristic bird-life of August
will be found in the marshes. There the Swallows

come in increasing numbers to their roosts in the reeds, while Red-winged Blackbirds, and Bobolinks under the alias of Reedbird, are abundant where the wild rice grows.

August is practically the last month of the nesting season as well as of the song season. The late-breeding Goldfinch and Waxwing are occupied with family matters in August, and Song Sparrows sometimes rear a third brood in this month; but with these exceptions, birds are rarely found nesting in August.

BIRDS OF THE MONTH.

PERMANENT RESIDENTS (see page 6.)

SUMMER RESIDENTS (see page 10).

MIGRANTS ARRIVING FROM THE NORTH.

Plate No.		Plate No.	
	August 1 to 15.		Tennessee Warbler.*
7.	Sora.*		Nashville Warbler.*
10.	Semipalmated Sandpiper.*		Parula Warbler.*
10.	Semipalmated Plover.*		Cape May Warbler.*
	Yellow-bellied Flycatcher.*	61.	Black-throated Green Warbler.*
	Golden-winged Warbler.*		
	Chestnut-sided Warbler.*		Black-throated Blue Warbler.*
	Canadian Warbler.*		
	Small-billed Water Thrush.*		Magnolia Warbler.*
			Blackburnian Warbler.*
	August 15 to 31.		Wilson's Warbler.*
	Olive-sided Flycatcher.*	71.	Red-breasted Nuthatch.†

* Transient Visitant passing further south.
† Irregular Winter Visitant.

SEPTEMBER.

The student whose patience has been sorely tried by the comparative scarcity of birds in August, will find that in September his observations in the field will be attended by far more interesting results. The first marked fall in the temperature is sure to be followed by a flight of migrants which, like the " waves " of May, will flood the woods with birds. The larger number will be Warblers; indeed, September, with May, is characterized by the abundance of these small birds.

Birds of the year will outnumber the adults, and, in most cases, their plumage will be quite unlike that worn by their parents in May, while, in many instances, even the adults themselves will appear in a changed costume. Often this new dress will resemble that of the immature bird, a fact which accounts for the apparent absence of old birds in the fall migration.

As a rule, fall plumages are less striking than those of spring, and when, in addition, it is remembered that birds are not in song, and that the foliage is much denser, the greater difficulty of field identification at this season will be appreciated.

In September more migrating birds are killed by striking lighthouses than in any other month of the year. This is doubtless owing to the fact that stormy or foggy weather is more apt to prevail in September than during any other period of active migration; that the majority of the migrants are young and inexperienced, and that in September migrants are more numerous than in any other month.

About September 25, our more common Winter Visitants arrive from the north, and after that date birds rapidly decrease in number.

Few songs are heard during the month; the characteristic bird-notes being the sharp *keè-yer* of the Flicker, and the calls of Blue Jays gathering their autumn toll from the chestnut trees.

BIRDS OF THE MONTH.

PERMANENT RESIDENTS (see page 6).

SUMMER RESIDENTS (see page 10).

The following will depart for the south:

Plate No.		Plate No.	
	September 1 to 10.		*September 20 to 30.*
	Acadian Flycatcher.	10.	Common Tern.
36.	Orchard Oriole.	6.	Little Green Heron.
	Rough-winged Swallow.	29.	Hummingbird.
	Worm-eating Warbler.	30.	Kingbird.
	Blue-winged Warbler.	31.	Crested Flycatcher.
		33.	Wood Pewee.
	September 10 to 20.	54.	Rose-breasted Grosbeak.
		59.	Yellow-throated Vireo.
35.	Baltimore Oriole.		Warbling Vireo.
86.	Purple Martin.		Hooded Warbler.
87.	Yellow Warbler.		Louisiana Water Thrush.
65.	Yellow-breasted Chat.	73.	Veery.

MIGRANTS ARRIVING FROM THE NORTH.

September 1 to 10.		Blue-headed Vireo.*
Lincoln's Sparrow.*		Olive-backed Thrush.*
Black-poll Warbler.*		Bicknell's Thrush.*
Connecticut Warbler.*		
		September 20 to 30.
September 10 to 20.		
		4. Herring Gull.†
9. Wilson's Snipe.*		5· Green-winged Teal.*

* Transient Visitant passing further south.
† Winter Visitant.

OCTOBER.

Early October generally brings the first killing frost, depriving insectivorous birds of a large part of their food, and of necessity forcing them to journey southward. Flycatchers, Warblers, Vireos, and Swallows now take their departure, and after the fifteenth of the month few insect-eating birds remain, except those which, like Woodpeckers, feed on insects' larvæ or eggs.

This is the season of Sparrows. In countless numbers they throng old stubble, potato, or corn fields, doing untold good by destroying the seeds of noxious weeds. Song, Field, Chipping, and Vesper Sparrows may be found in flocks, all harvesting the year's crop of seeds, and with them will be the lately arrived Juncos, Tree and Fox Sparrows. When disturbed, they seek shelter in the nearest hedgerow, and their mingled notes produce a twittering chorus, in which it is difficult to distinguish the voices of individual birds.

This, however, will not be the only bird music of the month. Certain species now have a brief

---

* Transient Visitant passing further south.
† Winter Visitant.

second song period, and on the brighter days of the month we may hear Song, White-throated, and Fox Sparrows, Phœbes, and Ruby-crowned Kinglets in song.

The diurnal migration of Crows and Hawks is a feature of the bird-life of the month. In scattered companies they string across the sky, *en route* to more productive feeding grounds.

BIRDS OF THE MONTH.

PERMANENT RESIDENTS (see page 6).

REMAINING SUMMER RESIDENTS (see page 10).

The following will depart for the south :

Plate No.	
October 1 to 10.	*October 10 to 20.*
6. Black-crowned Night Heron.	11. Spotted Sandpiper.
22. Yellow-billed Cuckoo.	27. Whip-poor-will.
Black-billed Cuckoo.	27. Nighthawk.
28. Chimney Swift.	59. Red-eyed Vireo.
Least Flycatcher.	64. Maryland Yellowthroat.
38. Bobolink.	69. Long-billed Marsh Wren.
Grasshopper Sparrow.	Short-billed Marsh Wren.
83. Indigo Bunting.	68. House Wren.
84. Scarlet Tanager.	67. Brown Thrasher.
1. Barn Swallow.	88. Catbird.
1. Cliff Swallow.	
1. Bank Swallow.	*October 20 to 31.*
White-eyed Vireo.	
60. Black and White Warbler.	2. Pied-billed Grebe.
62. Redstart.	32. Phœbe.
63. Oven-bird.	55. Towhee.
74. Wood Thrush.	1. Tree Swallow.

MIGRANTS ARRIVING FROM THE NORTH.

Plate
No.

October 1 to 10.

3 Loon.*
5. Pintail.*
5. Mallard.*
5. Canada Goose.*
Bronzed Grackle.*
Rusty Blackbird.*
American Pipit.
75. Hermit Thrush.

Plate
No.

October 10 to 20.

47. Fox Sparrow.*

October 20 to 31.

34. Horned Lark.†
Pine Finch.†
49. Tree Sparrow.†
50. Snowflake.†
50. Redpoll.†
58. Northern Shrike.†

NOVEMBER.

It is an interesting fact that the last migrants to leave in the fall are the first to arrive in the spring.

The bird-life of November, when the fall migration is practically concluded, closely resembles, therefore, that of March, when the spring migration is inaugurated.

The reason for this similarity is to be found in the fact that both months furnish birds with essentially the same kind of food. Thus the Loon, Grebes, Ducks, Geese, and Kingfisher remain until November or early December, when the forming of ice deprives them of food and forces them to seek open water. Woodcock and Snipe linger until they can no longer probe the frost-hardened earth; but the thaws of March will bring all these birds back to us by restoring their food.

---

* Transient Visitant passing further south.
† Winter Visitant.

Certain Sparrows stay with us until the weeds bearing the seeds on which they feed are covered by snow, when they are compelled to retreat further southward, returning, however, as soon as March suns lay bare the earth.

Few birds' songs are heard in November. In some sheltered, sun-warmed hollow, Song and White-throated Sparrows may continue in voice, but the characteristic bird-note of the month is the sweet, minor " scatter-call " of Bob-whites, who, after their sudden flight from the sportsman, endeavor to find one another by a questioning, whistled *where-are-you? where-are-you?*

BIRDS OF THE MONTH.

PERMANENT RESIDENTS (see page 6).

REMAINING SUMMER RESIDENTS (see page 10).

The following leave for the south, concluding the fall migration:

Plate No.		Plate No.	
5.	Wood Duck.	37.	Purple Grackle.
6.	Great Blue Heron.	40.	Cowbird.
7.	American Bittern.	44.	Vesper Sparrow.
76.	Woodcock.	43.	Field Sparrow.
13.	Mourning Dove.	45.	Chipping Sparrow.
23.	Belted Kingfisher.	42.	Swamp Sparrow.
82.	Red-winged Blackbird.		

DECEMBER.

The character of the bird-life of December depends largely upon the mildness or severity of the season. Should the ponds and streams remain open, the

ground be unfrozen, and little or no snow fall,
many of the migrant species of November will
linger into December. They rarely are found,
however, after the middle of the month, when our
bird-life is reduced to its simplest terms, being composed only of the ever-present Permanent Residents
and the Winter Visitants.

The comparative scarcity of food now forces birds
to forage actively for provisions, and when a supply
is found they are apt to remain until it is exhausted.
Their wanderings lead them over large areas, and
our dooryards and orchards may often be visited
by species which, when food is more abundant, do
not leave their woodland haunts. An excellent
way in which to attract them is to provide them
with suitable food. Crumbs and seeds scattered in
some place where they will not be covered by snow,
or blown away, will bring Juncos, Tree Sparrows,
and Purple Finches ; an old seed-filled sunflower
head may prove a feast for Goldfinches, while bits
of meat, suet, or ham bone hung from trees will be
eagerly welcomed by Chickadees, Nuthatches, and
Downy Woodpeckers.

LISTS OF BIRDS

THE dates given in the preceding review of the bird-life of the year will not, of course, hold good for localities far removed from the vicinity of New York city. Notes from various localities on the birds included in "Bird-Life" are, therefore, appended as a guide to students living in other parts of the eastern United States. These notes have been generously contributed by ornithologists whose long-continued observations make them the authorities on the birds of the sections from which they write.*

NOTES FROM WASHINGTON, D. C., ON BIRDS INCLUDED IN "BIRD-LIFE."

By Dr. C. W. Richmond.

Pied-billed Grebe.—Common Winter Visitant, August 25 to April or May.

Loon.—Common Winter Visitant, September to April 25.

Herring Gull.—Common Winter Visitant, October to March.

Common Tern.—Irregular Transient Visitant, sometimes common.

Wood Duck.—Uncommon Permanent Resident.

---

* The dates given in the following lists of birds are the average dates on which the species occur.

Pintail.—Winter Visitant, October to April.

Mallard.—Common Winter Visitant.

Green-winged Teal.—Common Winter Visitant, September to April.

Blue-winged Teal.—Common Winter Visitant, September to April.

Canada Goose.—Winter Visitant, and rather common Transient Visitant, October to April.

Great Blue Heron.—Rather common, absent only in midwinter.

Little Green Heron.—Very common Summer Resident, April 15 to September.

American Bittern.—Not uncommon Summer Resident; occasional in winter.

Sora.—Common Transient Visitant, March; July to November.

Clapper Rail.—Accidental; one record.

American Coot.—Common Transient Visitant, March to May; September to October 15.

Woodcock.—Rather common from February to November; a few winter.

Spotted Sandpiper.—Common Transient Visitant; not common Summer Resident, April 5 to September 30.

Wilson's Snipe.—Common Transient Visitant, March to May; September to November; occasional in winter.

Semipalmated Sandpiper.—Rare Transient Visitant, May; August to October.

Killdeer.—Permanent Resident, most abundant in migrations.

Semipalmated Plover.—Casual, three specimens, May; August.

Bob-white.—Common Permanent Resident.

Ruffed Grouse.—Not common Permanent Resident.

Mourning Dove.—Permanent Resident, common except in winter.

Turkey Vulture.—Abundant Permanent Resident.

Red-shouldered Hawk.—Common Permanent Resident.

Red-tailed Hawk.—Common Winter Visitant; rare Summer Resident.

Marsh Hawk.—Common Winter Visitant, July to April.

Sparrow Hawk.—Common Winter Visitant, rare Summer Resident.

Sharp-shinned Hawk.—Common Permanent Resident.

Cooper's Hawk.—Common Permanent Resident.

Bald Eagle.—Not common Permanent Resident.

Osprey.—Uncommon Summer Resident, March 25 to October.

Short-eared Owl.—Common Winter Visitant.

Long-eared Owl.—Common Permanent Resident.

Screech Owl.—Common Permanent Resident.

Barred Owl.—Not common Permanent Resident.

Yellow-billed Cuckoo.—Common Summer Resident, May 2 to October 15.

Black-billed Cuckoo.—Rather rare Summer Resident, May 2 to October 15.

Kingfisher.—Common Permanent Resident.

Downy Woodpecker.—Common Permanent Resident.

Hairy Woodpecker.—Rare Permanent Resident.

Red-headed Woodpecker.—Rather common Summer Resident; rare Winter Visitant.

Flicker.—Common Summer Resident; rare Winter Visitant.

Nighthawk.—Not common Summer Resident; abundant Transient Visitant, April 20 to October.

Whip-poor-will.—Common Summer Resident, April 15 to October.

Swift.—Abundant Summer Resident, April 15 to October 10.

Ruby-throated Hummingbird.—Common Summer Resident, April 28 to September.

Kingbird.—Common Summer Resident, April 20 to September.

Crested Flycatcher.—Very common Summer Resident, April 25 to September.

Phœbe.—Common Summer Resident, March 5 to October. Occasionally winters.

Least Flycatcher.—Common Transient Visitant, April 25 to May 25 ; August 28 to September 25.

Wood Pewee.—Common Summer Resident, April 28 to October 15.

Horned Lark.—Common Winter Visitant, November to March or April.

Crow.—Abundant Permanent Resident.

Blue Jay.—Rather rare Permanent Resident; common Transient Visitant, April 28 to May 15; September 15 to October 15.

Baltimore Oriole.—Rather common Summer Resident, April 28 to September.

Orchard Oriole.—Common Summer Resident, April 28 to September.

Red-winged Blackbird.—Common Permanent Resident, abundant in migrations.

Purple Grackle.—Common Transient Visitant and Summer Resident, February 20 ; a few winter.

Bobolink.—Transient Visitant, common in spring, abundant in fall, May 1 to 27 ; August 5 to October 1.

Meadowlark.—Common Permanent Resident; less common in winter.

Cowbird.—Rather rare Permanent Resident; common Transient Visitant.

Song Sparrow.—Common Permanent Resident; abundant Transient Visitant, March and October.

Swamp Sparrow.—Very common Transient Visitant, April to May 15; September 25 to October 30; a few winter.

Field Sparrow.—Very common Permanent Resident.

Vesper Sparrow.—Permanent Resident, very common in migrations; less so in summer and winter.

Chipping Sparrow.—Common Summer Resident; abundant Transient Visitant, March 15 to November 1; occasionally winters.

White-throated Sparrow.—Very common Winter Visitant, September 28 to May 20.

White-crowned Sparrow.—Irregularly common Winter Visitant and Transient Visitant, April 15 to May 15; October 15 to November 30.

Fox Sparrow.—Very abundant Transient Visitant, February 5 to April 5; October 25 to November; a few winter.

Junco.—Abundant Winter Visitant, October 5 to April 25.

Tree Sparrow.—Abundant Winter Visitant, November 1 to April 5.

Redpoll.—Very rare and irregular Winter Visitant.

Snowflake.—Casual in winter; one instance.

American Crossbill.—Irregular Winter Visitant, sometimes abundant.

Pine Grosbeak.—Casual in winter.

Goldfinch.—Common Permanent Visitant.

Purple Finch.—Common Winter Visitant, September 15 to May 15; largely a migrant.

Rose-breasted Grosbeak.—Rather common Transient Visitant, May 1 to 20; August 25 to October 1.

Towhee.—Common Summer Resident; very common Transient Visitant, April 15 to May 15; September to October 15 ; a few winter.

Indigo Bunting.—Common Summer Resident, April 28 to October 15.

Cardinal.—Common Permanent Resident ; less common than formerly.

Scarlet Tanager.—Common Transient Visitant ; rare Summer Resident, April 28 to October 7.

Barn Swallow.—Common Summer Resident ; more abundant Transient Visitant, March 28 to September.

Cliff Swallow.—Common Summer Resident ; more common Transient Visitant, April 15 to September 21.

Bank Swallow.—Rare Summer Resident, April to September.

Tree Swallow.—Common Transient Visitant, April 1 to May 25 ; July 10 to September.

Purple Martin.—Rather common Summer Resident, April 12 to September 15.

Cedar Waxwing.—Very common Permanent Resident ; less so in winter.

Northern Shrike.—Rare and irregular Winter Visitant, November to February.

Red-eyed Vireo.—Very common Summer Resident, April 25 to October 15.

Warbling Vireo.—Rather common Summer Resident, April 28 to September 10.

Yellow-throated Vireo.—Common Summer Resident, April 20 to September 15.

White-eyed Vireo.—Common Summer Resident, April 20 to October 7.

Black and White Warbler.—Abundant Transient Visitant ; less common Summer Resident, April 12 to October 15.

Yellow Warbler.—Common Summer Resident ; abundant Transient Visitant, April 18 to September 30.

Myrtle Warbler.—Abundant Winter Visitant, September 28 to May 20.

Black-throated Green Warbler.—Very common Transient Visitant, April 25 to May 28 ; August 28 to October 20.

Redstart.—Very abundant Transient Visitant, April 18 to May 28 ; August 19 to September 15.

Oven-bird.—Very common Summer Resident, April 20 to October 15.

Maryland Yellowthroat.—Abundant Summer Resident, April 18 to October 20.

Chat.—Common Summer Resident, April 29 to September.

Catbird.—Abundant Summer Resident, April 20 to October ; occasionally winters.

Mockingbird.—Uncommon Permanent Resident ; less numerous in winter.

Brown Thrasher.—Very common Summer Resident, April 5 to October 15 ; occasionally winters.

House Wren.—Common Summer Resident, April 15 to September.

Winter Wren.—Rather common Winter Visitant, September 25 to May.

Carolina Wren.—Common Permanent Resident.

Long-billed Marsh Wren.—Very numerous Summer Resident, April 30 to October.

Brown Creeper.—Common Winter Visitant, September 25 to April 25.

Carolina Chickadee.—Very common Permanent Resident, particularly in winter.

Tufted Titmouse.—Very common Permanent Resident ; more so in winter.

White-breasted Nuthatch.—Common Transient Visitant, and Winter Visitant; less common Summer Resident.

Red breasted Nuthatch.—Irregularly abundant Winter Visitant, sometimes rare, September 15 to May 10.

Golden-crowned Kinglet.—Abundant Winter Visitant, October 5 to April 27.

Ruby-crowned Kinglet.—Abundant Transient Visitant, April 5 to May 10 ; September 25 to November 1 ; occasionally winters.

Veery.—Common Transient Visitant, April 26 to May 28 ; August 20 to September 30.

Wood Thrush.—Common Summer Resident, April 20 to October 15.

Hermit Thrush.—Very common Transient Visitant ; sometimes not uncommon Winter Visitant, April 4 to May 15 ; October 15 to November.

Robin.—Rather common Summer Resident ; abundant Transient Visitant from February to April ; irregularly common in winter.

Bluebird.—Common Permanent Resident.

NOTES FROM A LOCALITY SLIGHTLY NORTH OF PHILADELPHIA, PA., ON THE BIRDS INCLUDED IN "BIRD-LIFE."

By Witmer Stone.

Pied-billed Grebe.—Common Transient Visitant.

Loon.—Tolerably common Transient Visitant and less frequent Winter Visitant, October 5 to May 1.

Herring Gull.—Common Winter Visitant, October 1 to April 1.

Common Tern.—Occasional in August.

Wood Duck.—Tolerably common Transient Visitant ; occasional Winter Visitant and Summer Resident.

Pintail.—Tolerably common Transient Visitant.

Mallard.—Not common Transient Visitant.

Green-winged Teal.—Tolerably common Transient Visitant.

Blue-winged Teal.—Common Transient Visitant.

Canada Goose —Common Transient Visitant, October 15 to April 15.

Great Blue Heron.—Tolerably common Summer Resident, April 1 to November 15; rare in winter.

Little Green Heron.—Common Summer Resident, April 1 to October 1.

Black-crowned Night Heron.—Common Summer Resident, April 15 to October 15; rare Winter Visitant.

American Bittern.—Tolerably Common Transient Visitant, April and September to November.

Sora.—Very common Transient Visitant, April and May, September and October.

Clapper Rail.—Very rare in Summer; very common Summer Resident at Atlantic City, N. J., April 15 to November 1.

Coot.—Not common Transient Visitant; occasional Winter Visitant.

Woodcock.—Formerly common Summer Resident, now rather rare and mainly Transient Visitant and occasional Winter Visitant.

Spotted Sandpiper.—Very common Summer Resident, April 20 to October 1.

Wilson's Snipe.—Tolerably common Transient Visitant, March 20 to May 10, and October, and occasional Winter Visitant.

Semipalmated Sandpiper.—Rare. Abundant Transient Visitant at Atlantic City, N. J., July 10 to October 1.

Killdeer.—Tolerably common Summer Resident; occasional Winter Visitant, March 20 to October 25.

Semipalmated Plover.—Rare. Common Transient Visitant at Atlantic City, N. J., May 10 to June 1 ; July 10 to September 15.

Bob-white. — Formerly common Permanent Resident ; scarcer in winter ; now becoming rare and mainly Transient Visitant.

Ruffed Grouse.—Formerly common Permanent Resident; now rare ; mainly in fall.

Mourning Dove.—Common Summer Resident and occasional Winter Visitant ; March to November.

Turkey Vulture.—Tolerably common Summer Resident ; occasional Winter Visitant.

Red-shouldered Hawk. — Tolerably common Permanent Resident.

Red-tailed Hawk.—Common Winter Visitant ; not common in summer.

Marsh Hawk. — Common Winter Visitant (rare Summer Resident ?)

Sparrow Hawk.—Common Permanent Resident.

Sharp-shinned Hawk.—Common Permanent Resident.

Cooper's Hawk.—Tolerably common Permanent Resident; very common Transient Visitant.

Bald Eagle.—Not Common Permanent Resident.

Osprey.—Tolerably common in Summer ; very common Summer Resident on New Jersey coast and Delaware Bay, March 20 to October 15.

Short-eared Owl.—Irregularly common Winter Visitant.

Long-eared Owl.—Not common Permanent Resident.

Screech Owl.—Very common Permanent Resident.

Barred Owl.—Rare; mostly in winter.

Yellow-billed Cuckoo.—Common Summer Resident, May 8 to October 1.

Black-billed Cuckoo.—Rare Summer Resident ; tolerably common Transient Visitant, May 8 to October 1.

Kingfisher.—Common Summer Resident; occasional Winter Visitant.

Downy Woodpecker.—Common Permanent Resident.

Hairy Woodpecker.—Rare; mainly in winter.

Red-headed Woodpecker.—Irregularly common Transient Visitant in fall ; tolerably common Summer Resident, but local ; occasional Winter Visitant.

Flicker.—Very common Summer Resident, March 25 to October 25 ; occasional during winter.

Nighthawk. — Common Transient Visitant; tolerably common, but rather local Summer Resident, May 4 to October 10.

Whip-poor-will.—Tolerably common Transient Visitant;

common Summer Resident in New Jersey, within twenty miles of Philadelphia; April 22 to September 30.

Swift.—Very common Summer Resident, April 15 to October 10.

Ruby-throated Hummingbird.—Common Summer Resident, May 7 to September 25.

Kingbird.—Common Summer Resident, May 1 to September 1.

Crested Flycatcher.—Common Summer Resident, May 1 to September 1.

Phœbe.—Common Summer Resident; occasional Winter Visitant, March 20 to October 25.

Least Flycatcher.—Tolerably common Transient Visitor, May 1 to 15; September 6 to 15.

Wood Pewee.—Common Summer Resident, May 6 to September 20.

Horned Lark.—Irregular Winter Visitant.

Crow.—Very common Permanent Resident.

Blue Jay.—Common Permanent Resident; less abundant in winter; most plentiful in fall.

Baltimore Oriole.—Tolerably common Summer Resident, May 1 to September 1.

Orchard Oriole.—Common Summer Resident, May 1 to September 1.

Red-winged Blackbird.—Common Summer Resident, February 20 to November; rather common Winter Visitant.

Purple Grackle.—Common Summer Resident, February 20 to November; occasional Winter Visitant.

Bobolink.—Tolerably common Transient Visitant, May 1 to 15; common Transient Visitant, August 25 to September 20.

Meadowlark.—Common Permanent Resident.

Cowbird.—Common Summer Resident, April 1 to October; occasional Winter Visitant.

Song Sparrow.—Abundant Permanent Resident.

Swamp Sparrow.—Tolerably common Permanent Resident; more abundant in migrations.

Field Sparrow.—Very common Summer Resident, March 18 to October; occasional Winter Visitant.

Vesper Sparrow.—Common Summer Resident, April 2 to November 1; occasional Winter Visitant.

Chipping Sparrow.—Common Summer Resident, March 30 to October 20.

White-throated Sparrow.—Very Common Transient Visitant, September 20 to May 20; Common Winter Visitant.

White-crowned Sparrow.—Rare Transient Visitant, May 2 to 13; October 6 to 20.

Fox Sparrow.—Very common Transient Visitant, March 10 to April 10; October 15 to December 1; occasional Winter Visitant.

Junco.—Very common Winter Visitant, October 1 to May 1.

Tree Sparrow.—Very common Winter Visitant, October 15 to April 15.

Redpoll.—Rare and irregular Winter Visitant.

Snowflake.—Rare and irregular Winter Visitant.

American Crossbill.—Rather rare and irregular Winter Visitant; has been seen in May.

Pine Grosbeak.—Only casual Winter Visitant.

Goldfinch.—Very common Permanent Resident.

Purple Finch.—Tolerably common Winter Visitant, September 25 to May 1; very common Transient Visitant.

Rose-breasted Grosbeak.—Tolerably common Transient Visitant, May 5 to May 12; September 5 to October 5.

Towhee.—Common Summer Resident, April 18 to October 20.

Indigo Bunting.—Common Summer Resident, May 10 to October 1.

Cardinal.—Tolerably common Permanent Resident.

Scarlet Tanager.—Common Transient Visitant, May 5 to May 18; September 10 to October 10; tolerably common Summer Resident.

Barn Swallow.—Common Summer Resident, April 14 to September 1.

Cliff Swallow.—Tolerably common Transient Visitant; rare Summer Resident, May 1 to September 1.

Bank Swallow.—Common but local Summer Resident, April 8 to April 20; September 1.

Tree Swallow.—Common Transient Visitant, April 20 to

May 15; August 15 to September 1; common Summer Resident in New Jersey, within twenty miles of Philadelphia.

Purple Martin.—Irregular and local Summer Resident, April 15 to September 1.

Cedar Waxwing.—Tolerably common Summer Resident; abundant Transient Visitant; occasional Winter Visitant.

Northern Shrike.—Rather rare Winter Visitant, December 2 to February 5.

Red-eyed Vireo.—Very common Summer Resident, April 30 to October 10.

Warbling Vireo.—Not very common Summer Resident, May 5 to October 10.

Yellow-throated Vireo.—Tolerably common Summer Resident, May 2 to September 15.

White-eyed Vireo.—Common Summer Resident, May 5 to October 1.

Black and White Warbler.—Very common Transient Visitant, April 23 to May 15 ; August 10 to October 5 ; less common Summer Resident.

Yellow Warbler.—Common Summer Resident, May 1 to September 25.

Myrtle Warbler.—Very common Transient Visitant, April 15 to May 20 ; September 25 to November 1 : found sparingly, Winter Visitant.

Black-throated Green Warbler.—Common Transient Visitant, May 1 to May 15 ; September 1 to October 10.

Redstart.—Very common Transient Visitant ; a few breed; April 30 to May 20 ; August 5 to October 5.

Oven-bird.—Common Summer Resident, April 30 to October 6.

Maryland Yellowthroat.—Very common Summer Resident, April 25 to October 12.

Chat.—Common Summer Resident, May 5 to September 20.

Catbird.—Very common Summer Resident, May 3 to October 18 ; one winter record.

Mockingbird.—Rare straggler.

Brown Thrasher.—Common Summer Resident, April 22 to October 20 ; occasional in winter ; a few records.

House Wren.—Tolerably common Summer Resident, April 25 to October 5.

Winter Wren.—Common Winter Visitant, September 25 to April.

Carolina Wren.—Tolerably common Permanent Resident.

Long-billed Marsh Wren.—Common Summer Resident ; a few winter.

Brown Creeper.—Common Transient Visitant ; less common Winter Visitant, September 20 to April 15.

Black-capped Chickadee.—Not common Winter Visitant, October 24 to March 1.

Tufted Titmouse.—Rather common Permanent Resident.

White-breasted Nuthatch.—Common Permanent Resident ; less numerous in summer.

Red-breasted Nuthatch.—Transient Visitant ; irregularly abundant in fall ; rare in spring ; May 15 to September 10 ; October 15 to May 15 ; and occasional Winter Visitant.

Golden-crowned Kinglet.—Very common Transient Visitant, September 30 to April 20 ; rather common Winter Visitant.

Ruby-crowned Kinglet.—Common Transient Visitant, April 12 to May 1 ; September 15 to November 1.

Veery.—Common Transient Visitant, May 5 to 25 ; September 1 to 20.

Wood Thrush.—Common Summer Resident, May 1 to October 1.

Hermit Thrush.—Very common Transient Visitant ; found sparingly as Winter Visitant ; April 10 to May 4 ; October 10 to November 5.

Robin.—Abundant Summer Resident ; frequent but irregular Winter Visitant ; March 15 to November 10.

Bluebird.—Tolerably common Transient Visitant ; rather rare Summer Resident ; formerly Permanent Resident. Beginning to increase again.

NOTES FROM PORTLAND, CONN., ON THE BIRDS INCLUDED IN "BIRD-LIFE."

By John H. Sage.

Pied-billed Grebe.—Transient Visitant, April 12 ; common, September 9 to November 22.

Loon.—Not common, Transient Visitant, April 21 ; September 25 to December 1.

Herring Gull.—Not common Winter Visitant, October 6 to March 8.

Wood Duck.—Common Transient Visitant, March 20 to April 8 ; September to December 3 ; a few breed.

Pintail.—Casual Transient Visitant, October 9 to 12.

Mallard.—Rare Transient Visitant, November 5.

Green-winged Teal.—Uncommon Transient Visitant, April 6 to 8 ; September to November 26.

Blue-winged Teal.—Uncommon Transient Visitant in fall, September 13 to October 20.

Canada Goose.—Common Transient Visitant, March 10 to May 8 ; October 13 to December 15.

Great Blue Heron.—Common Transient Visitant, April 3 to May 11 ; August 21 to November 25.

Little Green Heron.—Common Summer Resident, April 28 to October 14.

Black-crowned Night Heron.—Uncommon Summer Resident, April 15 to October 14.

American Bittern.—Not common Summer Resident, April 8 to October 24.

Sora.—Summer Resident, April to November 5.

Coot.—Transient Visitant, rare in April and May ; common, September 19 to November 14.

Woodcock.—Common Summer Resident, February 25 to November 28.

Spotted Sandpiper.—Common Summer Resident, April 22 to September 29.

Wilson's Snipe.—Common Transient Visitant, March 18 to

May 10 ; September 14 to November 30 ; one instance of breeding.

Semipalmated Sandpiper.—Common Transient Visitant in fall, August to October 7.

Killdeer.—Accidental Visitant, not seen since April 5, 1875.

Semipalmated Plover.—Transient Visitant, May 22 to June 4 ; September.

Bob-white.—Common Permanent Resident.

Ruffed Grouse.—Common Permanent Resident.

Mourning Dove.—Common Summer Resident, March 19 to November 30 ; occasional in winter.

Red-shouldered Hawk.—Common Permanent Resident ; less common in winter.

Red-tailed Hawk.—Common Permanent Resident ; less common in winter.

Marsh Hawk.—Tolerably common Summer Resident, April 1 to November 2.

Sparrow Hawk.—Rather rare Permanent Resident.

Sharp-shinned Hawk.—Common Summer Resident, March 27 to October 29 ; a few winter.

Cooper's Hawk.—Common Summer Resident, March 18 to October 15.

Bald Eagle.—Of irregular occurrence, April, May, June, and December.

Osprey.—Common Transient Visitant, April 5 to May 25 ; September 7 to October 18.

Short-eared Owl.—Common Transient Visitant, April ; October 8 to December 7.

Long-eared Owl.—Probably a Permanent Resident. Not uncommon in winter.

Screech Owl.—Common Permanent Resident.

Barred Owl.—Rare Permanent Resident, sometimes common in October, November, and December.

Yellow-billed Cuckoo.—Common Summer Resident, May 7 to October 17.

Black-billed Cuckoo.—Common Summer Resident, May 1 to September 4.

Kingfisher.—Common Summer Resident, April 5 to November 22; casual in winter.

Downy Woodpecker.—Common Permanent Resident.

Hairy Woodpecker.—Rare Permanent Resident.

Red-headed Woodpecker.—Rare Permanent Resident, sometimes common in fall; September 20 to November 28.

Flicker.—Common Summer Resident, March 8 to November 26; a few winter.

Nighthawk.—Common Summer Resident, April 28 to October 3.

Whip-poor-will.—Common Summer Resident, April 23 to September 25.

Swift.—Abundant Summer Resident, April 19 to October 11.

Ruby-throated Hummingbird.—Common Summer Resident, May 6 to September 22.

Kingbird.—Common Summer Resident, April 28 to September 10.

Crested Flycatcher.—Common Summer Resident, May 4 to August.

Phœbe.—Common Summer Resident, March 7 to October 14.

Least Flycatcher.—Common Summer Resident, April 21 to September 4.

Wood Pewee.—Common Summer Resident, May 6 to October 3.

Horned Lark.—Casual; March 22 to 25; no fall record.

Crow.—Common Permanent Resident.

Blue Jay.—Common Permanent Resident.

Baltimore Oriole.—Common Summer Resident, May 1 to September 8.

Orchard Oriole.—Summer Resident, May 10 to August.

Red-winged Blackbird.—Common Summer Resident, March 3 to November 1.

Bronzed Grackle.—Common Summer Resident, February 20 to November 8.

Bobolink.—Common Summer Resident, May 3 to October 15.

Meadowlark.—Common Summer Resident, March 8 to October 26; a few winter.

Cowbird.—Common Summer Resident, March 26 to November 6; occasional in winter.

Song Sparrow.—Permanent Resident; more common from March 5 to November 2.

Swamp Sparrow.—Not common Summer Resident, April 12 to November 2.

Field Sparrow.—Common Summer Resident, April 6 to October 26; occasional in winter.

Vesper Sparrow.—Common Summer Resident, April 4 to October 21.

Chipping Sparrow.—Abundant Summer Resident, April 5 to October 23.

White-throated Sparrow.—Very common Transient Visitant, April 13 to May 21; September 18 to November 12; occasional in winter.

White-crowned Sparrow.—Uncommon Transient Visitant, May 12 to 22; October 2 to 16.

Fox Sparrow.—Abundant Transient Visitant, March 2 to April 26; October 17 to November 27.

Junco.—Common Winter Visitant, September 28 to April 26.

Tree Sparrow.—Common Winter Visitant, October 26 to April 23.

Redpoll.—Irregular Winter Visitant, November 27 to March 31.

Snowflake.—Rather common Winter Visitant, October 25 to March 20.

American Crossbill.—Irregular in occurrence, December 10 to April 19.

Pine Grosbeak.—Irregular Winter Visitant, November 25 to March 25.

Goldfinch.—Common Permanent Resident.

Purple Finch.—Rather rare Permanent Resident; common Transient Visitant; irregular, but sometimes common in winter.

Rose-breasted Grosbeak.—Common Summer Resident, May 1 to September 28.

Towhee.—Common Summer Resident, April 27 to October 24.

Indigo Bunting.—Common Summer Resident, May 6 to October 16.

Scarlet Tanager.—Common Summer Resident, May 3 to October 7.

Barn Swallow.—Common Summer Resident, April 16 to October 19.

Cliff Swallow.—Summer Resident, less common than formerly, May 12 to September 14.

Bank Swallow.—Common Summer Resident, April 17 to September 25.

Tree Swallow.—Summer Resident, but common only as a migrant, April 5 to October 26.

Purple Martin.—Locally common Summer Resident, April 16 to September 12.

Cedar Waxwing.—Common Permanent Resident.

Northern Shrike.—Tolerably common Winter Visitant, October 26 to April 3.

Red-eyed Vireo.—Abundant Summer Resident, April 30 to October 8.

Warbling Vireo.—Common Summer Resident, April 29 to September 17.

Yellow-throated Vireo.—Common Summer Resident, April 26 to September 21.

White-eyed Vireo.—Common Summer Resident, May 3 to September 20.

Black and White Warbler.—Common Summer Resident, April 27 to October 6.

Yellow Warbler.—Common Summer Resident, April 29 to September 23.

Myrtle Warbler.—Common Transient Visitant, April 22 to May 19; September 21 to October 25; a few winter.

Black-throated Green Warbler.—Tolerably common Summer Resident, April 27 to October 21.

Redstart.—Common Summer Resident, May 2 to September 26.

Oven-bird.—Abundant Summer Resident, May 4 to September 26.

Maryland Yellowthroat.—Abundant Summer Resident, May 3 to November 7.

Chat.—Common Summer Resident, May 6 to August.

Catbird.—Common Summer Resident, April 30 to October 14.

Brown Thrasher.—Common Summer Resident, April 22 to October 20.

House Wren.—Tolerably common Summer Resident, April 21 to September 26.

Winter Wren.—Rather common Winter Visitant, September 23 to March 12.

Carolina Wren.—Accidental Visitant, March.

Long-billed Marsh Wren.—Locally abundant Summer Resident, May 18 to October 26.

Brown Creeper.—Tolerably common Winter Visitant, October 2 to May.

Black-capped Chickadee.—Common Permanent Resident.

White-breasted Nuthatch.—Tolerably common Permanent Resident.

Red-breasted Nuthatch.—Irregular Winter Visitant, September 18 to May 11.

Golden-crowned Kinglet.—Common Winter Visitant, October 8 to April 25.

Ruby-crowned Kinglet.—Common Transient Visitant, April 8 to May 6; September 26 to October 26.

Veery.—Common Summer Resident, May 3 to August 30.

Wood Thrush.—Common Summer Resident, May 4 to September 18.

Hermit Thrush.—Common Transient Visitant, April 6 to May 3; October 15 to 26; occasional in winter.

Robin.—Common Summer Resident, February 15 to November 21; a few winter.

Bluebird.—Common Permanent Resident.

NOTES FROM CAMBRIDGE, MASS., ON BIRDS INCLUDED IN "BIRD-LIFE."

By William Brewster.

Pied-billed Grebe.—Common in April; very common September to November; breeds in one locality.

Loon.—Not common Transient Visitant, April to early May; September to November.

Herring Gull.—Abundant Winter Visitant, November to April.

Common Tern.—Casual in September.

Wood Duck.—Common Transient Visitant, March and April; August to November; a few breed.

Pintail.—Casual Transient Visitant, April, September and October.

Green-winged Teal.—Uncommon Transient Visitant, April; September to November.

Blue-winged Teal. — Rare in spring ; very common, at least formerly, August to October.

Canada Goose.—Common Transient Visitant, March and April; October to December.

Great Blue Heron.—Common Transient Visitant, April and May; September to November ; occasional in summer.

Little Green Heron.—Common Summer Resident, May 5 to September.

Black-crowned Night Heron.—Permanent Resident, most common in August and September.

American Bittern.—Not common Summer Resident, April 15 to November.

Sora.—Very common Summer Resident, April 20 to October 20.

Coot.—Transient Visitant, rare in April; common September to November.

Woodcock.—Summer Resident, formerly common, fast becoming rare; March to November.

Spotted Sandpiper.—Common Summer Resident, April 26 to September.

Wilson's Snipe.—Common Transient Visitant, April 5 to May 5; September and October.

Semipalmated Sandpiper.—Very common in August and September.

Killdeer.—Accidental Visitant; two instances.

Semipalmated Plover.—Rare in spring ; sometimes common in August and September.

Bob-white.—Common Permanent Resident.

Ruffed Grouse.—Common Permanent Resident.

Mourning Dove.—Occasional during summer in immediate vicinity of Cambridge.

Red-shouldered Hawk.—Common Permanent Resident; less common in winter.

Red-tailed Hawk.—Common Winter Visitant, November to April ; a few in summer.

Marsh Hawk.—Common Transient Visitant, March 15 to April 15 ; September and October ; a few breed.

Sparrow Hawk.—Rather common Summer Resident, February to November.

Sharp-shinned Hawk.—Uncommon Transient Visitant, April 15 to April 30 ; September and October ; rare Summer Resident ; uncommon Winter Visitant.

Cooper's Hawk.—Common Transient Visitant, April, September, and October ; not common Summer Resident ; rare Winter Visitant.

Bald Eagle.—Of irregular occurrence at all seasons.

Osprey.—Common Transient Visitant, April ; September.

Short-eared Owl.—Uncommon Transient Visitant, April, October, and November.

Long-eared Owl.—Not common Permanent Resident.

Screech Owl.—Common Permanent Resident.

Barred Owl.—Rare Permanent Resident, sometimes common in November and December.

Yellow-billed Cuckoo.—Common Summer Resident, May 12 to August.

Black-billed Cuckoo.—Common Summer Resident, May 15 to September 20.

Kingfisher.—Common Summer Resident, April 10 to October.

Downy Woodpecker.—Common Permanent Resident.

Hairy Woodpecker.—Uncommon Winter Visitant.

Red-headed Woodpecker.—Irregular at all seasons ; sometimes common in fall.

Flicker.—Very common Summer Resident; common Winter Visitant.

Nighthawk.—Not uncommon Summer Resident, May 15 to September 25.

Whip-poor-will.—Common Summer Resident, April 28 to September 20.

Swift.—Abundant Summer Resident, April 25 to September 20.

Ruby-throated Hummingbird.—Uncommon Summer Resident, May 12 to September.

Kingbird.—Abundant Summer Resident, May 5 to September 1.

Crested Flycatcher.—Uncommon Summer Resident, May 15 to August.

Phœbe.—Common Summer Resident, March 25 to October 10.

Least Flycatcher.—Abundant Summer Resident, May 1 to August 25.

Wood Pewee.—Common Summer Resident, May 18 to September 10.

Horned Lark.—Common Transient Visitant, October 25 to November 25 ; March 25 to April 5.

Crow.—Abundant Permanent Resident.

Blue Jay.—Common Permanent Resident ; abundant Transient Visitant, April and May ; September and October.

Baltimore Oriole.—Very common Summer Resident. May 8 through August.

Orchard Oriole.—Summer Resident, sometimes rather common May 15 to July.

Red-winged Blackbird.—Abundant Summer Resident, March to August; a few winter.

Bronzed Grackle.—Abundant Summer Resident, March to October ; occasional in winter.

Bobolink.—Very common Summer Resident, May 8 to September 10.

Meadowlark.—Common Summer Resident ; not common Winter Visitant.

Cowbird.—Very common Summer Resident, May 8 to September 10.

Song Sparrow.—Very abundant Summer Resident, March 10 to November 1 ; locally common Winter Visitant.

Swamp Sparrow.—Abundant Summer Resident, April 12 to November 10; a few winter.

Field Sparrow.—Common Summer Resident, April 15 to November 1.

Vesper Sparrow.—Very common Summer Resident, April 5 to October 15.

Chipping Sparrow.—Abundant Summer Resident, April 15 to October 25.

White-throated Sparrow.—Very common Transient Visitant, April 25 to May 15; October 1 to November 10; a few winter.

White-crowned Sparrow.—Uncommon Transient Visitant, May 12 to 22; October 1 to 20.

Fox Sparrow.—Abundant Transient Visitant, March 15 to April 20 ; October 20 to November 15.

Junco.—Rather common Winter Visitant; abundant Transient Visitant, September 20 to November 25; March 20 to April 20.

Tree Sparrow.—Common Winter Visitant; abundant Transient Visitant, October 25 to November 25 ; March 20 to April 20.

Redpoll.—Irregular Winter Visitant, often very abundant, October 25 to April 10.

Snowflake.—Common Winter Visitant, October 25 to March 25; abundant in migrations.

American Crossbill.—Of irregular occurrence at all seasons.

Pine Grosbeak.—Irregular Winter Visitant, frequently common; sometimes abundant November to March.

Goldfinch.—Very common Permanent Resident.

Purple Finch.—Permanent Resident, very common from March to October ; irregular, but at times abundant in winter.

Rose-breasted Grosbeak.—Common Summer Resident, May 10 to September 10.

Towhee.—Common Summer Resident, April 25 to October 15.

Indigo Bunting.—Rather common Summer Resident, May 15 to September 25.

Cardinal.—Casual ; two instances.

Scarlet Tanager.—Rather common Summer Resident, May 12 to October 1.

Barn Swallow.—Common Summer Resident, but fast decreasing, April 20 to September 10.

Cliff Swallow.—Summer Resident, much less common than formerly, April 28 to September 1.

Bank Swallow.—Common Summer Resident, April 28 to September 1.

Tree Swallow.—Summer Resident, formerly common, now common only as a migrant, April 5 to October 12.

Purple Martin.—Locally common Summer Resident, April 20 to August 25.

Cedar Waxwing.—Not common Permanent Resident; common Summer Resident; abundant Transient Visitant, in spring. February 1 to April 25.

Northern Shrike.—Common Winter Visitant, November 1 to April 1.

Red-eyed Vireo.—Abundant Summer Resident, May 10 to September 10.

Warbling Vireo.—Common Summer Resident, May 10 to September 25.

Yellow-throated Vireo.—Common Summer Resident, May 8 to September 1.

White-eyed Vireo.—Rather rare Summer Resident, May 8 to September 20 ; formerly common.

Black and White Warbler.—Very common Summer Resident, April 25 to September 5.

Yellow Warbler.—Abundant Summer Resident, May 1 to September 30.

Myrtle Warbler.—Abundant Transient Visitant, April 18 to May 20 ; September 20 to November 3 ; a few winter.

Black-throated Green Warbler.—Very common Summer Resident, May 1 to October 15.

Redstart.—Very common Summer Resident, May 5 to September 20.

Oven-bird.—Abundant Summer Resident, May 6 to September 15.

Maryland Yellow-throat.—Abundant Summer Resident, May 5 to October 20.

Chat.—Rather rare Summer Resident, May 15 to (?).

Catbird.—Abundant Summer Resident, May 6 to September 30.

Mockingbird.—Rare Summer Resident, March to November.

Brown Thrasher.—Very common Summer Resident, April 25 to October 15.

House Wren.—Locally common Summer Resident, May 1 to September 25.

Winter Wren.—Transient Visitant, rather common, September 20 to November 25 ; rare, April 10 to May 1 ; a very few winter.

Long-billed Marsh Wren.—Locally abundant Summer Resident, May 15 to October ; sometimes a few winter.

Brown Creeper.—Common Transient Visitant, rather common Winter Visitant, September 25 to May 1.

Black-capped Chickadee.—Very common Permanent Resident ; more numerous in fall and winter.

White-breasted Nuthatch.—Permanent Resident, rare in summer, uncommon in winter ; common in migrations.

Golden-crowned Kinglet.—Very common Transient Visitant ; common Winter Visitant, September 20 to April 25.

Ruby-crowned Kinglet.—Rather common Transient Visitant, April 10 to May 5 ; October 10 to November 5.

Veery.—Very common Summer Resident, May 10 to September 8.

Wood Thrush.—Rather common Summer Resident, May 12 to September 15.

Hermit Thrush.—Very common Transient Visitant, April 16 to May 5 ; October 5 to November 15 ; occasionally one or two may winter.

Robin.—Very abundant Summer Resident ; irregular Winter Visitant.

Bluebird.—Common Summer Resident, March 6 to November 1 ; more numerous during March and November.

NOTES FROM THE NEIGHBORHOOD OF ST. LOUIS, MO., INCLUDING PARTS OF ST. LOUIS AND ST. CHARLES COUNTIES, ON THE BIRDS INCLUDED IN " BIRD-LIFE."

By Otto Widmann.

Pied-billed Grebe.—Common Transient Visitant ; rare Summer Resident, April 1 to December 1.

Loon.—Rare Transient Visitant, April and October.

Herring Gull.—Transient Visitant and Winter Resident, less common than formerly, September 20 to May 5.

Common Tern.—Rare Transient Visitant, May and September.

Wood Duck.—Breeds frequently ; common in migrations in February and March ; September and October.

Pintail.—Abundant Transient Visitant, February 13 to April 15; October 10 to December 1.

Mallard.—Abundant Transient Visitant and frequent Winter Resident, September 15 to April 25.

Green-winged Teal.—Abundant Transient Visitant and occasional Winter Resident, February 15 to April 25 ; October 1 to December 15.

Blue-winged Teal.—Abundant Transient Visitant, September 1 to December 1.

Canada Goose.—Abundant Winter Visitant, October 15 to April 1.

Great Blue Heron.—Common Summer Resident, April 1 to November 1.

Little Green Heron.—Common Summer Resident, April 10 to October 10.

Black-crowned Night Heron.—Tolerably common Summer Resident, April 10 to Oct. 10.

American Bittern.—Rather rare Summer Resident, April 10 to October 20.

Sora.—Tolerably common Summer Resident and very common Transient Visitant, April 10 to November 1.

American Coot.—Tolerably common Summer Resident and very common Transient Visitant, April 1 to November 1.

Woodcock.—Common Summer Resident, March 1 to November 15.

Spotted Sandpiper.—Common Summer Resident, April 15 to October 15.

Wilson's Snipe.—Common Transient Visitant, March 1 to May 1 ; September 6 to November 20.

Semipalmated Sandpiper.—Irregular Transient Visitant, May ; August 4 to October 17.

Killdeer.—Common Transient Visitant; infrequent Summer Resident, March 10 to November 15.

Semipalmated Plover.—Tolerably common Transient Visitant, April 26 to May 5 ; August 20 to September 17.

Bob-white.—Abundant Permanent Resident.

Ruffed Grouse.—Permanent Resident in hilly region south of St. Louis.

Mourning Dove.—Abundant Summer Resident; rare Winter Resident, March 10 to November 1.

Turkey Vulture.—Common Summer Resident, February 25 to November 1.

Red-shouldered Hawk.—Common Permanent Resident.

Red-tailed Hawk.—Common Permanent Resident, most numerous in fall and early winter.

Marsh Hawk.—Common Permanent Resident.

Sparrow Hawk.—Common Permanent Resident.

Sharp-shinned Hawk.—Fairly common Transient Visitant, February, March ; October to December.

Cooper's Hawk.—Rather rare Summer Resident; Transient Visitant, more common in fall, September 15 to November 1 ; February 15 to March 15; sometimes winters.

Bald Eagle.—Winter Resident, becoming scarce, September 1 to April 1.

Osprey.—Rather common Summer Resident, April 1 to October 1.

Short-eared Owl.—Tolerably common Winter Visitant, October 8 to April 1.

Long-eared Owl.—Not common Winter Visitant.

Screech Owl.—Common Permanent Resident.

Barred Owl.—Common Permanent Resident.

Yellow-billed Cuckoo.—Common Summer Resident, April 28 to October 23.

Black-billed Cuckoo.—Rare Summer Resident; fairly common Transient Visitant, May 1 to October 15.

Kingfisher.—Common Summer Resident, March 1 to November 1.

Downy Woodpecker.—Common Permanent Resident.

Hairy Woodpecker.—Fairly common Permanent Resident.

Red-headed Woodpecker.—Common Summer Resident and frequent Winter Resident, April 15 to October 1.

Flicker.—Common Summer Resident and frequent Winter Resident, May 15 to October 15.

Nighthawk.—Common Transient Visitant and tolerably common Summer Resident, April 25 to October 13. Bulk of Transient Visitants, May 5 to 25 ; August 25 to September 15.

Whip-poor-will.—Common Summer Resident, April 8 to October 10.

Chimney Swift.—Abundant Summer Resident, April 1 to October 20.

Ruby-throated Hummingbird.—Common Summer Resident, April 25 to October 20.

Kingbird.—Common Summer Resident, April 10 to September 1.

Crested Flycatcher.—Common Summer Resident, April 20 to September 1.

Phœbe.—Summer Resident, less common than formerly, March 1 to November 1.

Least Flycatcher.—Fairly common Transient Visitant, April 28 to May 15 ; September 1 to October 15.

Wood Pewee.—Common Summer Resident, April 28 to October 1.

Prairie Horned Lark.—Common Permanent Resident.

American Crow.—Common Permanent Resident ; abundant Winter Resident.

Blue Jay.—Abundant Permanent Resident.

Baltimore Oriole.—Common Summer Resident, April 20 to September 10.

Orchard Oriole.—Common Summer Resident, April 20 to September 1.

Red-winged Blackbird.— Common Summer Resident; abundant Transient Visitant; frequent Winter Resident, March 1 to May 15 ; September 15 to November 15.

Bronzed Grackle.—Abundant Summer Resident and Transient Visitant; rare Winter Resident, March 10 to May 1; October 1 to November 15.

Bobolink.—Tolerably common Transient Visitant, April 28 to May 28; August 20 to September 24.

Meadowlark.—Common Summer Resident; rare Winter Resident, March 10 to November 1.

Cowbird.—Common Summer Resident; rare Winter Resident; abundant Transient Visitant, March 10 to May 1; September 15 to November 1.

Song Sparrow.—Common Transient Visitant; fairly common Winter Resident; rare Summer Resident; March 10 to April 15; September 20 to November 10.

Swamp Sparrow.—Common Transient Visitant; rare Winter Resident, March 15 to May 15; September 20 to November 10.

Field Sparrow.—Common Summer Resident, March 10 to November 1.

Vesper Sparrow.—Tolerably common Transient Visitant, March 25 to April 10; October 15 to November 1.

Chipping Sparrow.—Common Summer Resident, March 15 to October 25.

White-throated Sparrow.—Abundant Transient Visitant; fairly common Winter Resident, March 10 to May 25; September 25 to November 10.

White-crowned Sparrow.—Rare Winter Resident; common Transient Visitant, April 20 to May 20; October 1 to November 1.

Fox Sparrow.—Common Transient Visitant; fairly common Winter Resident, March 10 to April 12; October 7 to November 10.

Junco.—Abundant Transient Visitant and very common

Winter Resident, March 10 to April 20; September 20 to November 15.

Tree Sparrow.—Common Winter Resident, November 1 to April 1.

Redpoll.—Rare Winter Visitant, January and February.

American Crossbill.—Rare Transient Visitant, February 22 to April 1; middle of November.

Goldfinch.—Common Permanent Resident; abundant Transient Visitant, March 10 to May 1; September 15 to October 10.

Purple Finch.—Common Winter Resident and abundant Transient Visitant, March 10 to May 1; September 15 to November 1.

Rose-breasted Grosbeak.—Common Summer Resident, April 25 to October 10.

Towhee.—Common Summer Resident; tolerably common Winter Resident; Transient Visitant, March 10 to April 15; September 25 to October 20.

Indigo Bunting.—Abundant Summer Resident, April 25 to October 10.

Cardinal.—Common Permanent Resident.

Lark Finch.—Fairly common Summer Resident, April 15 to September 1.

Dickcissel.—Abundant Summer Resident, April 15 to October 1.

Scarlet Tanager.—Common Summer Resident, April 20 to September 24.

Barn Swallow.—Not common Summer Resident, April 10 to October 10.

Cliff Swallow.—Common Summer Resident; abundant Transient Visitant, April 20 to September 24.

Bank Swallow.—Common Summer Resident; abundant Transient Visitant, April 20 to September 24.

Tree Swallow.—Common Transient Visitant; rare Summer Resident, March 15 to April 20; September 1 to October 20.

Purple Martin.—Common Summer Resident; abundant Transient Visitant, March 20 to September 24.

Cedar Waxwing.—Permanent Resident and breeder.

Baltimore Oriole.—Common Summer Resident, April 20 to September 10.

Orchard Oriole.—Common Summer Resident, April 20 to September 1.

Red-winged Blackbird.—Common Summer Resident; abundant Transient Visitant: frequent Winter Resident, March 1 to May 1; September 15 to November 15.

Bronzed Grackle.—Abundant Summer Resident and Transient Visitant; rare Winter Resident. March 10 to May 1; October to November 15.

Bobolink.—Tolerably common Transient Visitant, April 28 to May 28; August 20 to September 24.

Meadowlark.—Common Summer Resident; rare Winter Resident, March 1 to November 1.

Cowbird.—Common Summer Resident; rare Winter Resident; abundant Transient Visitant, March 10 to May September 15 to November 1.

Song Sparrow. - Common Transient Visitant; fairly common Winter Resident; rare Summer Resident; March to April 15; September 20 to November 10.

Swamp Sparrow.—Common Transient Visitant; rare Winter Resident. March 15 to May 15; September November 10.

Field Sparrow.—Common Summer Resident, March November I.

Vesper Sparrow.—Tolerably common Transient Visitant March 25 to April 1; October 15 to November 1.

Chipping Sparrow.— common Summer Resident, N to October 25.

White-throated Sparrow.—Abundant Transient fairly common Winter Resident, March 10 September 25 to November 10.

White-crowned Sparrow.—Rare Winter Resident Transient Visitant April 20 to May 20; O November 1.

Fox Sparrow.—Common Transient Visitant; common Winter Resident, March 10 to April to November 10.

Junco.—Abundant Transient Visitant and

Winter Resident, March 10 to Ap
November 15.

Tree Sparrow.—Common Winter
April 1.

Redpoll.—Rare Winter Visitant

American Crossbill.—Rare Tran
to April 1; middle of Novem

Goldfinch.—Common Perma
Transient Visitant, March
October 10.

Purple Finch.—Common W.
Transient Visitant, March
November 1.

Rose-breasted Grosbeak —
25 to October 10.

Towhee.—Common Sum
Winter Resident; Tr.
15; September 25 to Oct

Indigo Bunting.—Abu
October 10.

Cardinal—Common Per

Lark Finch.—Fairly
to September 1.

Dickcissel—Abundant
ber 1.

Scarlet Tanager.
September 24.

Barn Swallow.
to October 10.

Cliff Swallow.
Transient Vi

Bank Swallow
Transient V

Tree Swallo
Resident
20.

Purple M
Transien

Cedar W

pped
cka-

less

ident.
sitant,

ansient
12 to

isitant ;
eptember

May 5 to

pril 15 to

Visitant,

on Summer
nt, March 1

lent and fre-
between Feb-
vember 10.

i

Northern Shrike.—Rare Winter Visitant, November 15 to March 1.

Red-eyed Vireo.—Common Summer Resident, April 17 to September 25.

Yellow-throated Vireo.—Common Summer Resident, April 13 to October 11.

White-eyed Vireo.—Common Summer Resident, April 15 to October 15.

Black and White Warbler.—Common Summer Resident, April 16 to September 29.

Yellow Warbler.—Summer Resident, April 20 to August 13.

Myrtle Warbler.—Abundant Transient Visitant; frequent Winter Resident, March 12 to May 12; September 17 to November 7.

Black-throated Green Warbler.—Common Transient Visitant, April 26 to May 15; August 31 to October 8.

Redstart.—Common Summer Resident, April 16 to September 25.

Oven-bird.—Common Summer Resident, April 12 to October 2.

Maryland Yellowthroat.—Common Summer Resident, April 14 to October 2.

Chat.—Common Summer Resident, April 19 to September 25.

Catbird.—Common Summer Resident, April 16 to October 7.

Mockingbird.—Rather rare Summer Resident, and Permanent Resident, March or April to October.

Brown Thrasher.—Common Summer Resident, March 25 to October 20.

House Wren.—Common Summer Resident, April 9 to October 4.

Winter Wren—Rather rare Transient Visitant, March 25 to April 15 ; October 1 to 15.

Carolina Wren.—Common Permanent Resident ; not so common as twenty years ago.

Long-billed Marsh Wren.—Fairly common Summer Resident, April 28 to October 28.

Brown Creeper.—Common Transient Visitant; rare Winter Visitant, March 10 to April 10 ; September 23 to November 4.

Chickadee.—Common Permanent Resident; Black-capped Chickadee north of Missouri River; Carolina Chickadee south of it.

Tufted Titmouse.—Common Permanent Resident ; less common than formerly.

White-breasted Nuthatch.—Common Permanent Resident.

Red-breasted Nuthatch.—Irregular Transient Visitant, April 25 to May 10 ; September 4 to January 15.

Golden-crowned Kinglet.—Tolerably common Transient Visitant; rather rare Winter Resident; March 12 to April 10 ; September 29 to November 1.

Ruby-crowned Kinglet.—Abundant Transient Visitant ; rare Winter Resident, April 1 to May 6 ; September 17 to October 20.

Veery.—Tolerably common Transient Visitant, May 5 to 14 ; September 1 to 20.

Wood Thrush.—Common Summer Resident, April 15 to September 24.

Hermit Thrush.—Tolerably common Transient Visitant, April 1 to 27; October 1 to 25.

Robin.—Abundant Transient Visitant; common Summer Resident; tolerably common Winter Resident, March 1 to November 10.

Bluebird.—Tolerably common Summer Resident and frequent Winter Resident; migrates chiefly between February 25 and March 15; October 1 and November 10.

NOTES FROM OBERLIN, O., ON BIRDS INCLUDED IN "BIRD-LIFE."

By Prof. Lynds Jones.

Pied-billed Grebe.—Uncommon Transient Visitant.

Loon.—Not common Transient Visitant, late March to late October.

Herring Gull.—Common Transient Visitant on Lake Erie, March to May; September to November.

Common Tern.—Sometimes common Transient Visitant.

Wood Duck.—Uncommon Summer Resident.

Pintail.—Common Transient Visitant.

Mallard.—Now uncommon Transient Visitant.

Green-winged Teal.—Rare Transient Visitant.

Blue-winged Teal.—Uncommon Transient Visitant.

Canada Goose.—Common Transient Visitant along the rivers.

Great Blue Heron.—Tolerably common Summer Resident, March 20 to September 15.

Little Green Heron.—Common Summer Resident, April 20 to November 13.

American Bittern.—Tolerably common Summer Resident, late March.

Sora.—Tolerably common Summer Resident.

American Coot.—Common Summer Resident along the rivers.

Woodcock.—Common Summer Resident, April to November.

Spotted Sandpiper.—Common Summer Resident, April 10 to September 15.

Wilson's Snipe.—Common Transient Visitant in spring, March 19 to April 28.

Semipalmated Sandpiper.—Uncommon Transient Visitant.

Killdeer.—Common Summer Resident, March 1 to November 20.

Semipalmated Plover.—Uncommon Transient Visitant.

Bob-white.—Not common Permanent Resident.

Ruffed Grouse.—Rare Permanent Resident.

Mourning Dove.—Abundant Summer Resident, late March to November; rare in winter.

Turkey Vulture.—Tolerably common Summer Resident, April 1 to September 15.

Red-shouldered Hawk.—Common Summer Resident, March to November ; rare in winter.

Red-tailed Hawk.—Not common Summer Resident, February 1 to December 15; rare in winter.

Marsh Hawk.—Uncommon Summer Resident.

Sparrow Hawk.—Common Summer Resident, April to October; rare in winter.

Sharp-shinned Hawk.—Rare Permanent Resident.

Cooper's Hawk.—Not common Summer Resident ; rare in winter.

Bald Eagle.—Rare; common at Sandusky, and fairly common along the lake shore.

Osprey.—Rare ; only seen along the lake shore.

Short-eared Owl.—Rare.

Long-eared Owl.—Tolerably common Permanent Resident.

Screech Owl.—Tolerably common Permanent Resident.

Barred Owl.—Uncommon Permanent Resident.

Yellow-billed Cuckoo.—Common Summer Resident, May 10 to September.

Black-billed Cuckoo. — Common Summer Resident, May 5 to September 10.

Kingfisher.—Tolerably common Summer Resident, April to October ; rare in winter.

Downy Woodpecker.—Common Permanent Resident.

Hairy Woodpecker.—Common Permanent Resident.

Red-headed Woodpecker.—Abundant Summer Resident, April 15 to September 15; rare in winter.

Flicker.—Abundant Summer Resident, March to November; rare in winter.

Nighthawk.—Very variable Summer Resident, May to September.

Whip-poor-will.—Common Summer Resident, along streams only, May to September.

Chimney Swift.—Abundant Summer Resident, in towns; April 15 to October 10.

Ruby-throated Hummingbird.—Not common Summer Resident, May 10 to September 10.

Kingbird.—Common Summer Resident, April 27 to August 10.

Crested Flycatcher.—Common Summer Resident, April 28 to October 1.

Phœbe.—Common Summer Resident; late March to October.

Least Flycatcher.—Common Transient Visitant in spring, April 27 to May 22.

Wood Pewee.—Abundant Summer Resident, May 1 to September 12.

Prairie Horned Lark.—Common Permanent Resident. On December 18, 1897, I found both *alpestris* and *praticola* in a flock of some one hundred and twenty-five.

Crow.—Common Summer Resident; late February to November ; rare in winter.

Blue Jay.—Common Permanent Resident.

Baltimore Oriole.—Common Summer Resident, April 25 to September 1.

Orchard Oriole.—Rare Summer Resident.

Red-winged Blackbird. — Common Summer Resident, March 10 to November.

Bronzed Grackle.—Abundant Summer Resident, March to November; rare in winter.

Bobolink.—Abundant Summer Resident, April 23 to September 15.

Meadowlark.—Abundant Summer Resident, March to November ; rare in winter.

Cowbird.—Abundant Summer Resident, late March to October.

Song Sparrow.—Abundant Summer Resident, March to November ; rare in winter.

Swamp Sparrow.—Rare Summer Resident, late April.

Field Sparrow.—Common Summer Resident, April 1 to October 20.

Vesper Sparrow.—Abundant Summer Resident, late March to early November.

Chipping Sparrow.—Common Summer Resident, early April to October.

White-throated Sparrow.—Common Transient Visitant, April 12 to May 12 ; October to November.

White-crowned Sparrow.—Common Transient Visitant, May 1 to 19 ; September 22 to October 10.

Fox Sparrow.—Tolerably common Transient Visitant, March 25 to April 20; October 20 to November 10.

Junco.—Common Transient Visitant, late March to May ; October to December.

Tree Sparrow.—Common Winter Visitant, October 20 to April.

Redpoll.—Rare Winter Visitant.

Snowflake.—Rare Winter Visitant.

American Crossbill.—Very irregular Winter Visitant.

Pine Grosbeak.—Rare Winter Visitant.

Goldfinch.—Common Permanent Resident ; abundant Summer Resident.

Purple Finch.—Tolerably common Winter Visitant, October to May.

Rose-breasted Grosbeak.—Tolerably common Summer Resident, May to September.

Towhee.—Common Summer Resident, March 26 to October 20.

Indigo Bunting.—Common Summer Resident, May to October.

Cardinal.—Tolerably common Permanent Resident ; mostly along rivers.

Dickcissel.—Variable Summer Resident, May to September.

Lark Finch.—Summer Resident, becoming common, April 28 to September 1.

Scarlet Tanager.—Common Summer Resident, May to September.

Barn Swallow.—Common Summer Resident, April 15 to September.

Cliff Swallow.—Common Summer Resident, April 15 to August 15.

Bank Swallow—Common Summer Resident, April 20 to August 15.

Tree Swallow.—Rare Summer Resident, April 15 to August 15.

Purple Martin.—Common Summer Resident, April 1 to September 1.

Cedar Waxwing.—Variable Summer Resident. When it nests it remains the whole year.

Northern Shrike.—Not common Winter Visitant, November to March.

Red-eyed Vireo.—Common Summer Resident, late April to late September.

Warbling Vireo.—Common Summer Resident, late April to late September.

Yellow-throated Vireo.—Tolerably common Summer Resident, May 1 to September 10.

Black and White Warbler.—Common Transient Visitant, late April to May 15; September 10 to 20.

Yellow Warbler.—Common Summer Resident, April 20 to August 1.

Myrtle Warbler.—Common Transient Visitant, April 15 to May 15; September to November.

Black-throated Green Warbler.—Common Transient Visitant, April 25 to May 15; September 10 to 20.

Redstart.—Common Summer Resident, late April to October.

Oven-bird.—Common Summer Resident, late April to August.

Maryland Yellowthroat.—Common Summer Resident, late April to September.

Chat.—Not common Summer Resident, May to August 20.

Catbird.—Abundant Summer Resident, April 25 to October.

Brown Thrasher.—Common Summer Resident, April 15 to October.

House Wren.—Common Summer Resident, April 15 to October.

Winter Wren.—Scarcely common Winter Visitant, November to May 17.

Long-billed Marsh Wren.—Not common Summer Resident.

Brown Creeper.—Not common Transient Visitant, late March to May; October.

Chickadee.—Common Permanent Resident.

Tufted Titmouse.—Common Permanent Resident.

White-breasted Nuthatch.—Common Permanent Resident.

Red-breasted Nuthatch.—Common Transient Visitant, April 1 to May 17; October.

Golden-crowned Kinglet.—Common Winter Visitant, September 25 to April 25.

Ruby-crowned Kinglet.—Common Transient Visitant, April 1 to May 5; September 25 to October 20.

Veery.—Not common Transient Visitant and Summer Resident, May to September.

Wood Thrush.—Common Summer Resident, late April to September.

Hermit Thrush.—Not common Transient Visitant, April 15 to May 1; October.

Robin.—Abundant Summer Resident, February 15 to November 25; a few usually winter.

Bluebird.—Common Summer Resident; early March to November.

NOTES FROM IN AND NEAR MILWAUKEE, WIS., ON THE BIRDS INCLUDED IN "BIRD-LIFE."

By H. Nehrling.

Pied-billed Grebe.—Tolerably common Summer Resident, April 10 to November 15.

Loon.—More or less common Summer Resident, April 1 to November 15; becoming scarcer.

Herring Gull.—Very abundant Winter Visitant, October 10 to May 5.

Wood Duck.—Very rare Summer Resident, March 20 to October 25.

Pintail.—Summer Resident, March 18 to October 10.

Mallard.—Summer Resident, March 17 to November 25, and later.

Green-winged Teal.—March 17 ; November 20.

Blue-winged Teal.—April 10 ; October 28.

Canada Goose.—March; September 20 to October 1; movements very irregular.

Great Blue Heron.—Common Summer Resident, April 1 to October 1.

Little Green Heron.—Summer Resident, May 1 to September 20.

American Bittern.—Summer Resident, April 18 to September 25.

Sora.—Common Summer Resident, April 26 to October 2.

American Coot.—Common Summer Resident, March 28 to September 20.

Woodcock.—Summer Resident, April 25 to October 5.

Wilson's Snipe.—Rare Summer Resident; common during migrations; April 15 to October 5.

Spotted Sandpiper.—Common Summer Resident, April 28 to September 25.

Semipalmated Sandpiper.—May 6.

Killdeer.—Very common Summer Resident, March to October 10 ; nests in West Park.

Semipalmated Plover.—May 23.

Bob-white.—Extinct in Wisconsin.

Ruffed Grouse.—Permanent Resident, once common, now very rare.

Mourning Dove.—Summer Resident, April 30 to October 10.

Turkey Buzzard.—Very rare.

Red-shouldered Hawk.—Summer Resident, March 20 to November 1.

Red-tailed Hawk.—Summer Resident, March 20 to November 5.

Marsh Hawk.—Summer Resident, March 18 to October 15.

Sparrow Hawk.—Rather common Summer Resident, March 18 to October 10.

Sharp-shinned Hawk.—Summer Resident, April 10 to October 1.

Cooper's Hawk.—Summer Resident, April 20 to October 10.

Osprey.—Common Summer Resident, April 1 to September 20.

Short-eared Owl.—Permanent Resident.

Long-eared Owl.—Permanent Resident.

Screech Owl.—Common Permanent Resident; nests in the city.

Black-billed Cuckoo.—Summer Resident, May 8 to September 1.

Yellow-billed Cuckoo.—Rather abundant Summer Resident, May 9 to September 2 ; nests in orchards in the city.

Kingfisher.—Common Summer Resident, April 19 to September 18.

Downy Woodpecker.—Permanent Resident.

Hairy Woodpecker.—Summer Resident, April 17 to October 1 ; probably winters.

Red-headed Woodpecker.—Very common Summer Resident, April 30 to September 20 ; nests in the city.

Flicker.—Summer Resident, April 13 to September 25 ; nests in the city.

Nighthawk.—Abundant Summer Resident, May 17 to August 25 ; nests on house-tops in the city.

Whip-poor-will.—Rather scarce Summer Resident, May 20 to August 31.

Chimney Swift.—Abundant Summer Resident, May 12 to September 10.

Ruby-throated Hummingbird.—Summer Resident, May 9 to October 6.

Kingbird.—Common Summer Resident, May 9 to August 15.

Crested Flycatcher.—Very rare Summer Resident, May 11 to August 15.

Phœbe.—Common Summer Resident, March 20 to September 30.

Wood Pewee.—Common Summer Resident, May 20 to August 31; nests in the city.

Prairie Horned Lark.—Common Permanent Resident.

Crow.—Permanent Resident.

Blue Jay.—Common Permanent Resident; nests in the city.

Baltimore Oriole.—Not numerous Summer Resident, May 9 to August 25.

Red-winged Blackbird. — Abundant Summer Resident, March 17 to November 1.

Bronzed Grackle.—Abundant Summer Resident, March 20 to October 26; nests in the city.

Bobolink.—Common Summer Resident, May 9 to September 1; not half so abundant as fifteen years ago.

Cowbird.—Very numerous Summer Resident, April 8 to September 15.

Meadowlark.—Common Summer Resident, March 17 to October 31.

Song Sparrow.—Common Summer Resident, March 17 to October 10; rare near the city.

Field Sparrow.—Rare Summer Resident, April 21 to September 20.

Chipping Sparrow.—Common Summer Resident, May 1 to September 10; nests in the city.

Vesper Sparrow.—Very common Summer Resident, April 10 to September 25.

White-throated Sparrow.—Very common Transient Visitant, April 28 to May 20; September 20 to October 22.

White-crowned Sparrow.—Abundant Transient Visitant, May 2 to 23; September 20 to October 4.

Fox Sparrow.—Irregular Transient Visitant, April 4 to April 18; October 25 to November 2.

Junco.—October 1 to April 10; breeds about 70 miles north of city.

Tree Sparrow.—November 1 to March 20; breeds farther north.

Redpoll.—Irregular Winter Visitant, occasionally abundant.

Snowflake.—Irregular Winter Visitant, sometimes abundant.

American Crossbill.—Irregular Winter Visitant, sometimes abundant.

Pine Grosbeak —Irregular Winter Visitant, sometimes abundant.

American Goldfinch.—Summer Resident, May 1 to October; occasional as late as December 25.

Purple Finch.—Summer Resident, April 12 to November 6; breeds sparingly.

Rose-breasted Grosbeak.—Rather common Summer Resident, May 9 to September 15; nests in the city.

Towhee.—Common Summer Resident, April 26 to September 15.

Indigo Bunting.—Rather common Summer Resident, May 9 to September 10.

Scarlet Tanager.—Summer Resident, May 9 to August 15; nests in the city.

Barn Swallow.—Common Summer Resident, April 25 to August 25.

Cliff Swallow.—Summer Resident, April 30 to August 26.

Bank Swallow.—April 30 to ——?

Tree Swallow.—Common Summer Resident, April 25 to August 25; nests in the city.

Purple Martin.—Common Summer Resident, April 26 to August 20.

Cedar Waxwing.—Permanent Resident of irregular movements; thousands winter; others migrate southward, returning in May.

Northern Shrike.—Winter Resident, November 1 to March 5.

Red-eyed Vireo.—Summer Resident, May 9 to August 25.

Warbling Vireo.—Summer Resident, May 11 to August 25.

Yellow-throated Vireo —Summer Resident, May 19 to August 20; nests in the city.

Black and White Warbler.—May 1 to August 27; breed farther north.

Yellow Warbler.—Common Summer Resident, May 9 to August 26; nests in the city.

Black-throated Green Warbler.—Common Transient Visitant, May 9 to May 15; August 15 to September 1.

Myrtle Warbler.—Transient Visitant, April 17 to May 1; October 1 to October 10.

Redstart.—Common Summer Resident, May 9 to August 26.

Oven-bird.—Abundant Summer Resident, May 6 to September 5.

Yellow-breasted Chat.—Rare.

Maryland Yellowthroat.—Common Summer Resident, May 5 to August 20.

Catbird.—Common Summer Resident, May 5 to August 25.

Brown Thrasher.—Summer Resident, April 25 to September 1.

House Wren.—Common Summer Resident, May 1 to August 25.

Long-billed Marsh Wren.—Common Summer Resident, May 15 to September 6.

Brown Creeper.—April 4 to October 28 ; appears to breed near here.

Chickadee.—Permanent Resident.

White-breasted Nuthatch.—Permanent Resident.

Red-breasted Nuthatch.—April 22, and again November 1 ; movements irregular ; breeds farther north.

Golden-crowned Kinglet.—April 4, and again October 1 to October 10.

Ruby-crowned Kinglet.—April 10, and again September 3 to October 2.

Veery.—Summer Resident, breeding sparingly, May 9 to August 31.

Wood Thrush.—Summer Resident, April 31 to September 2.

Hermit Thrush.—April 10 ; October 1 ; breeds farther north.

Robin.—Common Summer Resident, March 17 to October 1.

Bluebird —Summer Resident, March 17 to October 15.

Factors of Evolution (Chapter II, pages 14–16).—Give examples illustrating the diversity shown in the structure and habits of birds. What theory has been advanced to account for the wide variation in structure shown by birds? What is meant by Natural Selection? How does the theory of Lamarck differ from that of Darwin? How may the tail-feathers of the Woodpecker have acquired their present pointed shape? Is it probable that the Woodpecker's barbed tongue has been acquired in the same manner?

FORM AND HABIT.

The Wing (Chapter II, pages 17–24).—Name the functions of the wing. What is doubtless its most primitive use as an organ of locomotion? How is it used by young Gallinules? By the young Hoatzin? How is it used by Grebes and Penguins? By the Ostrich? What variation in expanse of wings is presented by birds? What relation exists between shape of wing and style of flight? Give illustrations. Mention some flightless birds. Why is flight necessary to the Razor-billed Auk? Under what conditions might it exist without the power of flight? What group of flightless birds is found in the Antarctic region? Where do they nest? Why? What birds become temporarily flightless? In what manner? What lake is inhabited by a flightless Grebe? Where are flightless Gallinules found? How did they probably reach the islands they now inhabit? Mention other flightless birds.

---

* The value of these lessons will be greatly enhanced if the teacher will constantly have the pupil name additional species in illustration of the facts and theories here mentioned.

In what manner is the wing sexually adorned ? How is it used as a musical instrument ? How may it express emotion ?

The Tail (Chapter II, pages 25-27).—Mention some birds in which the tail is sexually developed. What is the tail's main office ? . Give illustrations of its relation to the character of flight. What birds use the tail as a prop ? Describe the tail of the Motmot. How may the tail express emotion ? Give illustrations.

The Feet (Chapter II, pages 27-30).—What relation exists between the feet and wings ? Give illustrations of the relation between the structure of the feet and the manner in which they are used. On what is length of foot sometimes dependent ? Describe the Jacana's toes ? Of what assistance are they to the bird ? What birds use the feet in scratching for food? What birds use the foot as a hand ? Of what special use is it to the Birds of Prey ? Mention several species which use the foot as a weapon. Describe the seasonal modification in the feet of Grouse.

The Bill (Chapter II, pages 30-34).—To what human organ does the bill correspond in use ? Mention some of the functions of the bill. What is its most important office ? What does the bill in effect become ? To what is its shape in Hummingbirds related ? Give illustrations. What is a marked character of the bill of some fish-eating birds ? How is the bill used by some shore birds ? Give illustrations. Describe the shape and uses of the Huia-bird's bill.

COLORS OF BIRDS.

Color and Age (Chapter III, page 36).—What is the character of the nest plumage of birds that run or swim at birth ? Of birds that are reared in a nest ? Give illustrations. What plumage follows the nest plumage ? Does it resemble that of the parent ? How long is it worn ? Does the immature plumage sometimes differ from that of the adult ? Give illustrations. When does the Bobolink acquire his full plumage ? When, the Orchard Oriole ?

Color and Season (Chapter III, page 37).—When the

male differs from the female, what seasonal change in color may occur? If the sexes are alike, is there much variation in color?

The Molt (Chapter III, pages 37, 38).—How are changes in a bird's plumage chiefly accomplished? Is the process of molting subject to much variation? What are these variations apparently dependent on? At what time of the year do all birds molt? What usually occurs the following spring? Do any birds have a complete spring molt? Are special plumes ever acquired at this season? Describe the manner in which the Snowflake gains its breeding dress.

Color and Food (Chapter III, page 39).—How is the color of Canaries sometimes altered? What is the effect of red pepper on fowls? What is sometimes fed to Parrots to change their color? How do Flamingoes and Scarlet Ibises illustrate the relation between color and food? What color does the Purple Finch become in captivity?

Color and Climate (Chapter III, pages 39–41).—How does climate affect the colors of birds? What does this demonstrate? How many races of Song Sparrows are known? What relation exists between their colors and the climate of the regions in which they live? Where are the extremes in color found? Are these extremes connected? What is the prevailing character of the colors of Arizona birds? Of northwest coast birds? What are these races of birds? Under what conditions might they become species?

Color and Haunt and Habit (Chapter III, pages 41–44).— What is necessary to an understanding of the value of the colors of birds? What is the office of protective coloring? What of deceptive coloring? What are the prevailing colors of ground-inhabiting birds? Give examples. Are tree-inhabiting birds brighter than those that live on the ground? What explanation is advanced to account for this? How do we receive an erroneous idea of the colors of tropical birds? What has Mr. Thayer proved? What fact does he call attention to? How does this tend to conceal the animal? How does Mr. Thayer demonstrate his theory ? Mention one of the best arguments for the value of protective coloration. Give illustrations. What birds illustrate the value of deceptive coloring ? What are recognition or signalling colors ? Give illustrations.

Color and Sex (Chapter III, pages 45-47).—The pupil should learn the Synopsis of Secondary Sexual Characters, and give one or more illustrations of each kind of sexual difference mentioned. Explain and illustrate Darwin's theory of sexual selection. How does the theory of Wallace differ from that of Darwin ?

THE MIGRATION OF BIRDS.

Extent of Migration (Chapter IV, page 49).—Upon what is the extent of migration often dependent ? Explain this. Where do most migratory western species winter ? Where do our eastern migratory Sparrows and berry eaters winter ? Where do the majority of our eastern insectivorous species winter ? What route do they follow ? How far south do some Plover and Snipe winter ?

Times of Migration (Chapter IV, pages 49-53).—This branch of the study of bird migration is covered much more fully under the section devoted to seasonal lessons, where the method of treatment is suggested. The matter here given should be used in connection with the added material in the section named.

Manner of Migration (Chapter IV, pages 54-57).—What is the first step in the fall migration ? Do old or young birds lead the way ? What birds fly by night ? Why ? Give examples. What birds migrate chiefly by day ? Why ? Give examples. What birds migrate exclusively by day ? Why? Give examples. What constitute highways of migration ? At what height may migrating birds travel ? Of what advantage is this height to them ? When are birds attracted to lighthouses ? How may one observe the night migration of birds ? How many birds were thus observed at Tenafly, N. J. ? Describe the observations made from the Bartholdi Statue.

Origin of Migration (Chapter IV, pages 58–61).—What theory is here advanced to account for the origin of bird migration? What other animals migrate? What do most animals seek during the period of reproduction? Give illustrations. Describe the migrations of certain sea birds. What has been the probable influence of the glacial period on bird migration? Describe the route followed by Bobolinks when migrating. What does this illustrate? In what manner does the migration of birds resemble the flight of the Carrier Pigeon?

THE VOICE OF BIRDS.

Song (Chapter V, page 62).—What is song? What is its chief function? Mention several types of bird music. To what does the song season correspond? When and by what species is it inaugurated? When is it practically concluded? Is there a second song period? What birds first cease singing? What birds are midsummer singers?

Call-notes (Chapter V, page 65).—What is the relation of call-notes to song? What do the calls of the Robin express? Do birds inherit the calls and songs? Do they ever acquire the notes of other species?

THE NESTING SEASON.

Time of Nesting (Chapter VI, page 64).—At what season do migratory birds nest? When do tropical birds nest? Why are birds obliged to nest at a certain season? Give some examples illustrating the relation between nesting time and food.

Mating (Chapter VI, page 65).—(See page 45, Synopsis of the Secondary Sexual Characters of Birds.)

The Nest (Chapter VI, pages 65–68).—What is the first step in nest-building? Mention several sites in which birds may nest. What is the chief desideratum? Why can sea birds often lay their eggs in exposed places? How is temperament shown in nesting? Mention several kinds of material used by birds in nest-building. How have birds as nest-builders been classified ?　Do both sexes assist in nest-building ?　How much time may be consumed in the construction of a nest ?　Mention the eight factors governing the character of birds' nests and give examples illustrating each.　(See Plates XXVI, XXVIII, XXIX, XCII to C.)

The Eggs (Chapter VI, pages 68–70).—How many eggs may compose a full set ?　If the nest is robbed, will the eggs be replaced ?　Give illustrations.　Of what is the eggshell composed ?　To what is the color of eggs due ?　How may variations in color be effected ?　Is there much variation in the color of the eggs of the same species ?　Why are the eggs of præcocial birds larger than those of altricial birds ?　Give examples.　What are the extremes in the period of incubation ?　Do both sexes incubate ?　(See Plates XCI to C.)

The Young (Chapter VI, page 70).—The mental and physical growth of the Chicken forms an excellent and practical lesson in the development of a young bird.　A newly hatched chick may be procured and placed in a suitable cage in the class-room, where its actions and plumage may be closely studied.　Experiments may be made, showing how little inherited knowledge the chick possesses, by giving it bits of worsted, etc., to eat, and observing how it learns what is and what is not edible, how it does not instinctively recognize water, etc., and at the same time notes should be kept of its changes in plumage.

CLASSIFICATION OF THE BIRDS OF NORTH AMERICA.*

ORDER I.—PYGOPODES (DIVING BIRDS).

Ducklike birds, with generally sharply pointed bills; feet webbed, placed far back near the tail; tarsus much flattened; hind toe, when present, with a lobe or flap; bill without toothlike projections; tail very short, and sometimes apparently wanting.

FAMILY 1.—*Podicipidæ.* Grebes; 6 species.
FAMILY 2.—*Urinatoridæ.* Loons; 5 species.
FAMILY 3.—*Alcidæ.* Auks, Murres, and Puffins; 22 species.

ORDER II.—LONGIPENNES (LONG-WINGED SWIMMERS).

Birds with sharply pointed and frequently hooked or hawklike bills; toes four (except in the genus *Rissa*), the front ones webbed: wings long and pointed.

FAMILY 4.—*Stercorariidæ.* Skuas and Jaegers; 4 species.
FAMILY 5.—*Laridæ.* Gulls and Terns; 43 species.
FAMILY 6.—*Rynchopidæ.* Skimmers; 1 species.

ORDER III.—TUBINARES (TUBE-NOSED SWIMMERS).

Bill hawklike, the tip of the upper mandible generally much enlarged; nostrils opening through tubes; hind toe

---

* The arrangement and nomenclature here given is based on the American Ornithologist Union's Check-List, 2d edition, 1895.

reduced to a mere nail, and sometimes entirely wanting ; front toes webbed.

FAMILY 7.—*Diomediidæ.* Albatrosses; 4 species.

FAMILY 8.—*Procellariidæ.* Fulmars, Petrels, and Shearwaters; 28 species.

ORDER IV.—STEGANOPODES (TOTIPALMATE SWIMMERS).

Toes four; all connected by webs.

FAMILY 9.—*Phaëthontidæ.* Tropic Birds ; 2 species.

FAMILY 10.—*Sulidæ.* Gannets; 6 species.

FAMILY 11.—*Anhingidæ.* Darters; 1 species.

FAMILY 12.—*Phalacrocoracidæ.* Cormorants; 6 species.

FAMILY 13.—*Pelecanidæ.* Pelicans ; 3 species.

FAMILY 14.—*Fregatidæ.* Man-o'-War Birds; 1 species.

ORDER V.—ANSERES (LAMELLIROSTRAL SWIMMERS).

Toes four, the front ones fully webbed ; tarsus not flattened as in the Grebes ; bill with toothlike projections, fluted ridges, or gutters on its sides.

FAMILY 15.—*Anatidæ.* Ducks, Geese, and Swans; 54 species.

ORDER VI.—ODONTOGLOSSÆ (LAMELLIROSTRAL GRALLATORES).

Toes four, the front three webbed ; bill with toothlike ridges as in some Ducks, the end half bent downward ; legs long ; tarsus 12·00 inches or more in length.

FAMILY 16.—*Phœnicopteridæ.* Flamingoes ; 1 species.

ORDER VII.—HERODIONES (HERONS, STORKS, IBISES, ETC.).

Toes four, all on the same level, slightly or not at all webbed ; lores bare ; legs and neck generally much lengthened.

FAMILY 17.—*Plataleidæ*. Spoonbills ; 1 species.

FAMILY 18.—*Ibididæ*. Ibises ; 4 species.

FAMILY 19.—*Ciconiidæ*. Storks and Wood Ibises ; 2 species.

FAMILY 20. — *Ardeidæ*. Herons, Bitterns, etc. ; 15 species.

ORDER VIII.—PALUDICOLÆ (CRANES, RAILS, ETC.).

Toes four; middle toe without a comb, generally not webbed ; hind toe generally small, higher than front ones, or when on the same level (Gallinules and Coots only), the bill is comparatively short and stout, and the forehead has a bare shield ; lores feathered, or with hair-like bristles (Cranes).

FAMILY 21.—*Gruidæ*. Cranes ; 3 species.

FAMILY 22.—*Aramidæ*. Courlans ; 1 species.

FAMILY 23.—*Rallidæ*. Rails, Gallinules, and Coots ; 17 species.

ORDER IX.—LIMICOLÆ (SHORE BIRDS).

Toes four or three ; the hind toe, when present, less than half the length of the inner one, and always elevated above the others ; legs generally long and slender, the lower half of the tibiæ bare ; bill, in the true Snipe, generally long, slender, and soft, the nostrils opening through slits or grooves ; wings long and pointed, the first primary generally the largest.

FAMILY 24.—*Phalaropodidæ*. Phalaropes; 3 species.

FAMILY 25.—*Recurvirostridæ*. Avocets and Stilts; 2 species.

FAMILY 26.—*Scolopacidæ*. Snipes and Sandpipers; 43 species.

FAMILY 27.—*Charadriidæ*. Plovers; 13 species.

FAMILY 28.—*Aphrizidæ*. Surf Birds and Turnstones; 3 species.

FAMILY 29.—*Hæmatopodidæ*. Oyster-catchers; 4 species.

FAMILY 30.—*Jacanidæ*. Jacanas; 1 species.

ORDER X.—GALLINÆ (GALLINACEOUS BIRDS).

Toes four, the hind one small and elevated above the front ones; bill generally short, stout, hard, and horny; wings short, the outer primaries curved and much stiffened.

> FAMILY 31.—*Tetraonidæ.* Grouse, Partridges, etc.; 20 species.
> FAMILY 32.—*Phasianidæ.* Pheasants, Turkeys, etc.; 1 species.
> FAMILY 33.—*Cracidæ.* Curassows, Guans, etc.; 1 species.

ORDER XI.—COLUMBÆ (PIGEONS).

All four toes on the same level; the hind toe about as long as the shortest front one; bill rather slender, deeply grooved; the nostrils opening in a soft fleshy membrane or skin.

> FAMILY 34.—*Columbidæ.* Pigeons; 13 species.

ORDER XII.—RAPTORES (VULTURES, HAWKS, AND OWLS).

All four toes armed with strong, sharp, curved nails or talons; the hind toe, except in the Vultures, as long as or longer than the shortest front one; bill with a cere, or covering of skin, at its base, through which the nostrils open, very strong and stout, the tip of the upper mandible with a sharply pointed hook.

> FAMILY 35.—*Cathartidæ.* American Vultures; 3 species.
> FAMILY 36.—*Falconidæ.* Vultures, Falcons, Hawks, Eagles, etc.; 39 species.
> FAMILY 37.—*Strigidæ.* Barn Owls; 1 species.
> FAMILY 38.—*Bubonidæ.* Horned Owls, Hoot Owls, etc.; 17 species.

ORDER XIII.—PSITTACI (PARROTS, PAROQUETS, ETC.).

Toes four, two in front and two behind ; bill with a cere, or covering of skin, at its base.

FAMILY 39. — *Psittacidæ*. Parrots and Paroquets ; 1 species.

ORDER XIV.—COCCYGES (CUCKOOS AND KING-FISHERS).

Toes four, two in front and two behind (Cuckoos), or three in front, the middle and outer ones joined for half their length; bill without a cere.

FAMILY 40.—*Cuculidæ*. Cuckoos, Anis, etc. ; 7 species.
FAMILY 41.—*Trogonidæ*. Trogons ; 1 species.
FAMILY 42.—*Alcedinidæ*. Kingfishers ; 3 species.

ORDER XV.—PICI (WOODPECKERS).

Toes four, or, rarely, three; two in front; bill strong; tail-feathers usually pointed and stiffened.

FAMILY 43.—*Picidæ*. Woodpeckers ; 24 species.

ORDER XVI.—MACROCHIRES (GOATSUCKERS, SWIFTS, AND HUMMINGBIRDS).

Feet very small and weak ; bill short, and mouth large (Goatsuckers and Swifts), or bill long and exceedingly slender (Hummingbirds); wings generally long and pointed.

FAMILY 44.—*Caprimulgidæ*. Goatsuckers ; 6 species.
FAMILY 45.—*Micropodidæ*. Swifts ; 4 species.
FAMILY 46.—*Trochilidæ*. Hummingbirds ; 18 species.

ORDER XVII.—PASSERES (PERCHING BIRDS).

Toes four, without webs, all on the same level ; hind toe as long as the middle one ; its nail generally longer than that of the middle one; foot, therefore, fitted for perching.

FAMILY 47.—*Cotingidæ.* Cotingas ; 1 species.

FAMILY 48.—*Tyrannidæ.* Flycatchers ; 33 species.

FAMILY 49.—*Alaudidæ.* Larks ; 2 species.

FAMILY 50.—*Corvidæ.* Crows, Jays, Magpies, etc.; 20 species.

FAMILY 51.—*Sturnidæ.* Starlings ; 1 species.

FAMILY 52.—*Icteridæ.* Blackbirds, Orioles, etc.; 20 species.

FAMILY 53.—*Fringillidæ.* Finches, Sparrows, etc.; 94 species.

FAMILY 54.—*Tanagridæ.* Tanagers ; 6 species.

FAMILY 55.—*Hirundinidæ.* Swallows ; 10 species.

FAMILY 56.—*Ampelidæ.* Waxwings, etc.; 3 species.

FAMILY 57.—*Laniidæ.* Shrikes ; 2 species.

FAMILY 58.—*Vireonidæ.* Vireos ; 12 species.

FAMILY 59.—*Cœrebidæ.* Honey Creepers ; 1 species.

FAMILY 60.—*Mniotiltidæ.* Wood Warblers; 59 species.

FAMILY 61.—*Motacillidæ.* Wagtails ; 7 species.

FAMILY 62.—*Cinclidæ.* Dippers ; 1 species.

FAMILY 63.—*Troglodytidæ.* Wrens, Thrashers, etc.; 25 species.

FAMILY 64.—*Certhiidæ.* Creepers; 1 species.

FAMILY 65.—*Paridæ.* Nuthatches and Tits; 21 species.

FAMILY 66. *Sylviidæ.* Kinglets and Gnatcatchers; 7 species.

FAMILY 67.—*Turdidæ.* Thrushes, Bluebirds, etc.; 15 species.

TEACHERS' MANUAL OF BIRD-LIFE,

TO ACCOMPANY PORTFOLIOS OF

FULL-PAGE COLORED PLATES.

Photographic bromide copies of the original drawings for " Bird-Life" have been carefully colored by an expert colorist, under the author's supervision, and are here reproduced by lithography.

Contains the same text as the "Teachers' Edition," and is sold only with the Portfolios of colored plates, as follows :

PORTFOLIO No. I.

PERMANENT RESIDENTS AND WINTER VISITANTS. 32 PLATES.

PORTFOLIO No. II.

MARCH AND APRIL MIGRANTS. 34 PLATES.

PORTFOLIO No. III.

MAY MIGRANTS, TYPES OF BIRDS' EGGS, AND NINE HALF-TONE PLATES SHOWING TYPES OF BIRDS' NESTS FROM PHOTOGRAPHS FROM NATURE. 34 PLATES.

Price of the Portfolios, each, $1.25; the Manual, 75 cents; the three Portfolios, with the Manual, $4.00.

D. APPLETON AND COMPANY, NEW YORK.

*B*IRD–LIFE. A Guide to the Study of our Common Birds. By FRANK M. CHAPMAN, Assistant Curator of Mammalogy and Ornithology, American Museum of Natural History ; Author of " Handbook of Birds of Eastern North America." With 75 full-page Plates and numerous Text Drawings by Ernest Seton Thompson. 12mo. Cloth, $1.75.

Also, edition in colors of the above, 8vo, cloth, $5.00.

" A volume exceptionally well adapted to the requirements of people who wish to study common birds in the simplest and most profitable manner possible. . . . As a readily intelligible and authoritative guide this manual has qualities that will commend it at once to the attention of the discerning student."—*Boston Beacon.*

" An interesting mass of data collected through years of study and observation. . . . While accurate from a scientific point of view, it makes delightful reading for those who will soon be among the flowers and the fields."—*Philadelphia Inquirer.*

" A careful reading of this book, which is well indexed, will open the eyes of many who have never seen the beauties of our birds before, and one can not help being interested in the book. While the ornithologists owe Mr Chapman a debt of gratitude for putting forth such a delightful volume, the ordinary reader owes him more, bringing, as he does, that reader in close touch with a new and beautiful world—the world of birds. The book is decidedly charming from every point of view."—*Cincinnati Commercial Tribune.*

" Unusually beautiful in itself, but it deserves praise because the colored pictures of the birds approach more nearly the natural appearance than usual. . . . Compared with these, the colored pictures of birds one usually sees are gaudy."—*Boston Herald.*

" His chronicles are full of the enthusiasm of the born naturalist. He gossips about the affairs of birds in a delightful strain, making ' Bird-Life ' an irresistible invitation to a fuller study of ornithology. It is not dry details he offers, but pretty stories, biographical sketches of interesting families—all sorts of birdlore, that proves the most enchanting reading. A great advantage in this work will be found in the beautifully colored illustrations, . . . which have received the greatest care in preparation."—*Chicago Evening Post.*

*H*ANDBOOK OF BIRDS OF EASTERN NORTH AMERICA. With Keys to the Species ; Descriptions of their Plumages, Nests, etc. ; their Distribution and Migrations. By FRANK M. CHAPMAN. With nearly 200 Illustrations. 12mo. Library Edition, cloth, $3.00 ; Pocket Edition, flexible morocco, $3.50.

" A book so free from technicalities as to be intelligible to a fourteen-year-old boy, and so convenient and full of original information as to be indispensable to the working ornithologist. . . . As a handbook of the birds of eastern North America it is bound to supersede all other works.' —*Science.*

" The author has succeeded in presenting to the reader clearly and vividly a vast amount of useful information."—*Philadelphia Press.*

" A valuable book, full of information compactly and conveniently arranged."—*New York Sun.*

" A charming book of interest to every naturalist or student of natural history."—*Cincinnati Times-Star.*

" The book will meet a want felt by nearly every bird observer."—*Minneapolis Tribune.*

D. APPLETON AND COMPANY, NEW YORK.

9 781330 531921